SOLAR FLARE PREDICTION

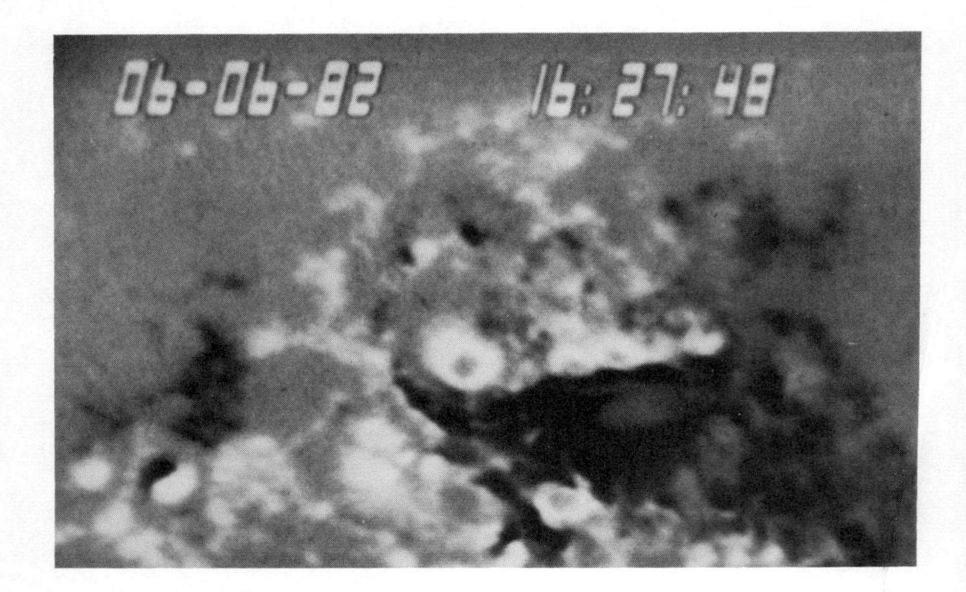

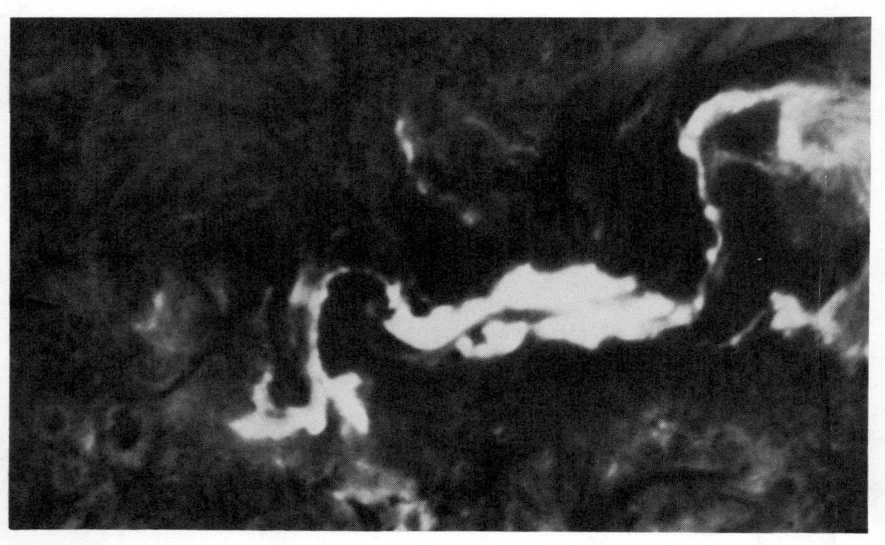

Frontispiece. Photospheric magnetic field (above) and the flare of 1982 June 6 at H-alpha - 0.6 A (below); a fine example of a two-ribbon flare that lies along the neutral line separating opposite polarities of the line-of-sight magnetic field. (Courtesy of Big Bear Observatory.)

SOLAR FLARE PREDICTION

C. Sawyer, J.W. Warwick, and J.T. Dennett
Department of Astrophysical, Planetary
and Atmospheric Sciences
University of Colorado, Boulder

COLORADO ASSOCIATED UNIVERSITY PRESS

Box 480, University of Colorado
Boulder, Colorado 80309
International Standard Book Number 0-87081-159-2
Library of Congress Catalog Card Number 85-073399
Printed in the United States of America

Colorado Associated University Press is a cooperative publishing enterprise supported in part by Adams State College, Colorado State University, Fort Lewis College, Mesa College, Metropolitan State College, University of Colorado, University of Northern Colorado, University of Southern Colorado, and Western State College

TABLE OF CONTENTS

LIST OF FIGURES

LIST OF TABLES

PREFACE

This started out as a limited project: to report on the status of solar-flare prediction. As we investigated various aspects of the problem and learned from various colleagues, the project grew. It could be argued that flare prediction represents the whole field of solar activity, for most activity is related eventually to flares, and we can't claim to understand flares until we can either predict them or define the stochasticism that may limit ability to predict them.

In order to keep growth within some bounds, we've adopted a narrow definition of flares themselves, and have not tried to include the equally fascinating, extensive, and practically important material on effects of flares on Earth and in space. We could not have made this separation a couple of decades ago when we relied on Earth's atmosphere, ionosphere, and magnetic field as detectors of solar radiation. We do it now with regret for the loss of tight cooperation between geophysics and solar physics. The need to understand the whole system has not disappeared.

There is no journal of flare prediction; most of what is written on the subject has appeared in proceedings of meetings and workshops, or as unpublished reports. In collecting and interpreting this material, the help of colleagues has been indispensable. They have proved generous and patient, as well as knowledgable and efficient. We can mention here only a few of those who have been particularly helpful in reading, criticizing, and suggesting improvements: George Dulk, Karen Harvey, Jo Ann Joselyn, Sara Martin, Donald Neidig, David Rust, and David Webb.

Gary Heckman, Joseph Hirman, Jesse Smith, and David Speich have generously instructed us in the practice of forecasting. Vic Gaizauskas, Jack Harvey, Gordon Hurford,

Gerald van Hoven, Allen Krieger, L. Krivsky, Margaret Liggett, Patrick McIntosh, Margaret Shea, Herschel Snodgrass, Peter Sturrock, and Harold Zirin were particularly helpful in providing additional material; Karl Kildahl in supplying data; Theodore Speiser in discussing magnetic reconnection; and Dominic Vecchia in discussing the importance of the data base in numerical forecasting. Joseph Crawford, John Cooper, and Steve Dominguez gave advice that was vital to getting words onto floppies. Final editing was carried out with vigilence and goodwill by Sandra Rush. The Committee on University Scholarly Publication of the University of Colorado, Boulder, aided with a grant for preparing illustrations. The work was supported by the Office of Naval Research, Electronics Division, through contract N00014-84-K-0131. We acknowledge with gratitude the support and encouragement of the sponsor, Lewis Larmore.

1. Introduction

1.1 SIGNIFICANCE OF FLARES AND THEIR EFFECTS

Flares on the sun were quite unsuspected before 1859. Today, although flares may be a source of wonder and delight, their effects are a source of concern to scientists, to industries, and to governments. Considerable effort goes into predicting the effects of flares on human activities on Earth and in space. Descriptions of the most significant effects of flares follow.

X-rays enhance the electron density in the ionosphere, disturbing radio communications. A radio wave propagating in the ionosphere is absorbed or refracted, depending upon the wave frequency. One result of the enhanced electron density can be immediate and sudden loss of short-wave radio communication that can last an hour or more. Another is an abrupt change in transmission path of navigational signals, which can lead to erroneous determinations of location.

Energetic protons damage electronic components of detectors, transducers, and computers in satellites. They endanger humans in space and in stratospheric flights over the poles, where protection of the geomagnetic field is diminished. They enhance the ionization density of the polar ionosphere, resulting in increased absorption of high-frequency waves propagated into that region.

The geomagnetic field can change abruptly due to a variation of the magnetic field and momentum of the solar wind impinging on Earth's magnetosphere. This disturbs the ionosphere, and affects measurements used in geophysical exploration. Induced currents can disrupt electric power transmission and cable communications and affect oil pipelines. Heating of the upper atmosphere associated with geomagnetic disturbances changes its density structure, increases atmospheric drag on satellites, and thwarts efforts to predict satellite orbits.

1.2 OVERVIEW

In preparing this report on solar-flare prediction we investigated two areas that are often kept separate: the observations, analyses, and theories aimed at understanding the flare process, and the practical operations of gathering and interpreting relevant data and making and distributing the flare forecast. A search of the literature included unpublished reports and proceedings of meetings and workshops as well as research papers and reviews in scholarly journals and books. We queried solar physicists and forecasters about the possibility of making useful flare predictions, and about efforts and plans for the future. Observatory and forecast-center visits were part of the investigation of the collection of solar data and its interpretation in the form of activity forecasts.

This study concentrates on the forecasting of flares themselves and barely touches on their geophysical consequences, although these are the bases of practical interest in flare prediction. The variation in fluxes of X-rays, energetic protons, and momentum, the propagation of charged particles and of shocks and structures in the solar wind, and the response of Earth's magnetosphere, ionosphere, and atmosphere are of tremendous scientific and practical interest, but beyond the scope of this report.

The literature review revealed statistical dependence of flare occurrence on active-region characteristics, and many suggestive precursors, but no distinctive set of solar phenomena consistently linked to certain flare occurrence in a specific time frame. The search also revealed that the abstracted journals seldom printed reports of research on flare prediction. Few papers published outside of workshop proceedings included either "forecast" or "prediction" in their titles or in their key words.

This contrasts with the rich assortment of reports of new observations and theories of flares: observations in parts of the spectrum that have not been seen before, observations with high spatial resolution that reveal new structure and complexity, and observations with high time resolution that show correspondingly fine bursts of emission, with each new observation sure to be quickly matched with at least one new theoretical explanation. Multispectral observations that coordinate information from a number of detectors are particularly clarifying. New basic theoretical insights give added confidence that solar physicists are on the track of understanding flare mechanisms, and are making rapid progress. This progress is reported in Chapter 2. Observations of possible flare precursors receive special emphasis in Chapter 3.

Despite rapid progress, flare physics is still far from attaining the exact and quantitative understanding of flare processes required for accurate prediction. On the other hand, most flares occur in active regions, and most big flares occur in strong, complex magnetic fields; when an endeavor is sensitive to flare occurrence, an experienced forecaster with current information on solar activity can offer a great deal of guidance. But how much can be expected?

To find out how solar physicists perceive forecast possibilities we devised a short questionnaire and distributed it to researchers, forecasters, and managers interested in flare prediction. Opinions varied widely. Fewer than half the responses expressed even a qualified belief that flares are statistically predictable. Many saw the possibility of predicting the location of a flare but few expressed much hope of predicting the time of the flare. Opinions about flare precursors were interesting, thoughtful, and highly individual. The people we queried held widely varying opinions about how well flares can be predicted and how to go about it.

At the 1984 Meudon solar-activity prediction workshop, various working groups summarized progress since the 1979 Boulder workshop. The only report that abandoned conventional optimism was that of the short-term solar prediction group. D. Neidig, co-chair with J. Smith, observed (partly for shock value, he explained later) that the five-year period had brought "no major improvements, maybe no minor improvements." He then went on to examine possible reasons for this apparent stagnation.

Neidig and Smith's concern contrasts sharply with the cheerful summaries of other working groups. For example, speaking for the medium-term prediction group, P. McIntosh said ". . . new computers and image processing have made some difference," and J. Joselyn reported that the working group on geomagnetic activity had learned a lot, found new techniques, and learned to "pose the questions better."

The question raised, of course, by the frank report of the short-term working group is how much improvement in forecast accuracy can we expect. We can anticipate improvement, in the long run, from better physical understanding of flares. The progress described in Chapters 2 and 3 provides a basis for this expectation. The experience of meteorological forecasting points to a possible source of more immediate progress through use of numerical guidance, with a computer to summarize current data in the framework of a large set of past data. Work on quantitative, objective prediction of solar flares is discussed in Chapter 4. Chapter 5 describes operational aspects of forecasting, shows that flare forecasting is only a small part of the total effort, and asks what forecasters need to make their jobs easier and to do their jobs better.

How good are the forecasts now? This is another question that lacks a clear answer at present. In Chapter 6 we look at scoring methods and how they apply to solar-flare forecasts, and we raise some questions about interpretation. The route for evaluating forecasts is not well marked but is at present a tangle of non-standardized terms, differing objectives, and reluctance to compare the skills of individual forecasters.

In the final chapter, we look at plans for future observational programs, and summarize and state conclusions: progress in forecasting flares may not follow immediately upon progress in physical understanding--research is an investment for the future; more immediate improvement of forecasts may be gained by use of numerical guidance, maintenance of an accurate data base, and realistic verification and evaluation of forecasts.

2. *Recent Progress in Understanding Flares*

2.1 INTRODUCTION

A solar flare is a system of complex processes that includes relaxation of stressed magnetic field, release of energy, and radiation of the energy. The classical flare, a "sudden, local brightening in H-alpha," is only a part of the total event--one of the few parts that can be seen in the narrow visible band of the wavelength spectrum. H-alpha flares appear in the chromosphere, usually in active regions with sunspots. Early studies described shape and size, rise and decay, and changes in spectral lines that, analyzed, yield temperature, density, and motion. Accompanying radio bursts were detected. Simultaneous enhancements of electron density in the ionosphere recorded bursts of ionizing radiation. Mass ejection was deduced from geomagnetic and ionospheric storms that followed some outstandingly large flares. "Solar cosmic-ray increases" were discovered and linked to flares. Soon after the International Geophysical Year in 1958 systematic observations of flares in H-alpha reached a peak with development of a global network of cinematographic stations using birefringent filters that isolate a narrow spectral range within the line profile to show flare structure with high contrast and sensitive photographic film to show flare development with high time resolution (Smith and Smith 1963; Svestka 1976).

Now bursts of radiation at short and long extremes of wavelength and enhanced fluxes of energetic charged particles are described as parts of the same system. This definition developed over recent decades as observatories were placed above the ionosphere and atmosphere and beyond the magnetosphere to detect the variable solar fluxes with improving spectral, spatial, and temporal resolution.

Bursts of X rays, gamma rays, and EUV radiation absorbed in Earth's atmosphere and radio bursts at frequencies too low to penetrate the ionosphere are now part of the picture. Energetic charged particles are observed beyond Earth's magnetospheric shield. With some flares, magnetized plasma is ejected into the corona, produces shock waves, and structures the magnetic field in the distant solar wind. A growing collection of flares have been observed from various viewpoints, clarifying their structure and their development; defining density, temperature, pressure, and magnetic field; and hinting at new details such as concentration of density in fine filaments and of energy in small populations of charged particles.

The data of Figure 2.1 range over 18 decades (almost 60 octaves) of frequency and wavelength; in the more familiar comparison of the solar wavelength spectrum to 6000-degree blackbody radiation, the information hardly extends beyond the upper tenth of the narrow band marked "visible." Well outside this band and, for biological creatures, comfortably below its peak, the two-million-degree corona and various kinds of solar activity become manifest. The logarithmic scales in Figure 2.1 and the chosen ordinate, the product of flux with frequency, offer advantages but introduce distortion that should be taken into consideration. The advantages of such scales, however, are a unified presentation and easy conversion to wavelength or energy units. Part of

---

Figure 2.1 (facing page)--The spectrum of solar radiation and its variation. The ordinate is the logarithm of the product of frequency (Hz) and radiative flux (ergs s^{-1} cm^{-2} Hz^{-1}). The abscissa is the logarithm of the frequency, or, alternatively, of wavelength, or of energy per quantum. On the low-frequency side, the upper dashed curve represents the envelope of radio bursts, and the similar lower curve represents typical large bursts. Solar-cycle and active-region variation is represented by enhanced width of the "no-flare" curve near centimeter wavelengths (log f ~ 10) and in the X-ray spectrum. The dashed curve on the right represents the upper envelope for X-ray bursts. The few gamma-ray flares for which spectra have been published are shown in more detail in Figure 2.2. Data sources are Malitson (1977), Cliver et al. (1982), Shimabukuro (1977), Steinberg et al. (1984), Steinberg et al. (1985), Tokar and Gurnett (1980), Weber (1978), Wild et al. (1963), Zieba et al. (1982), and sources listed with Figure 2.2.

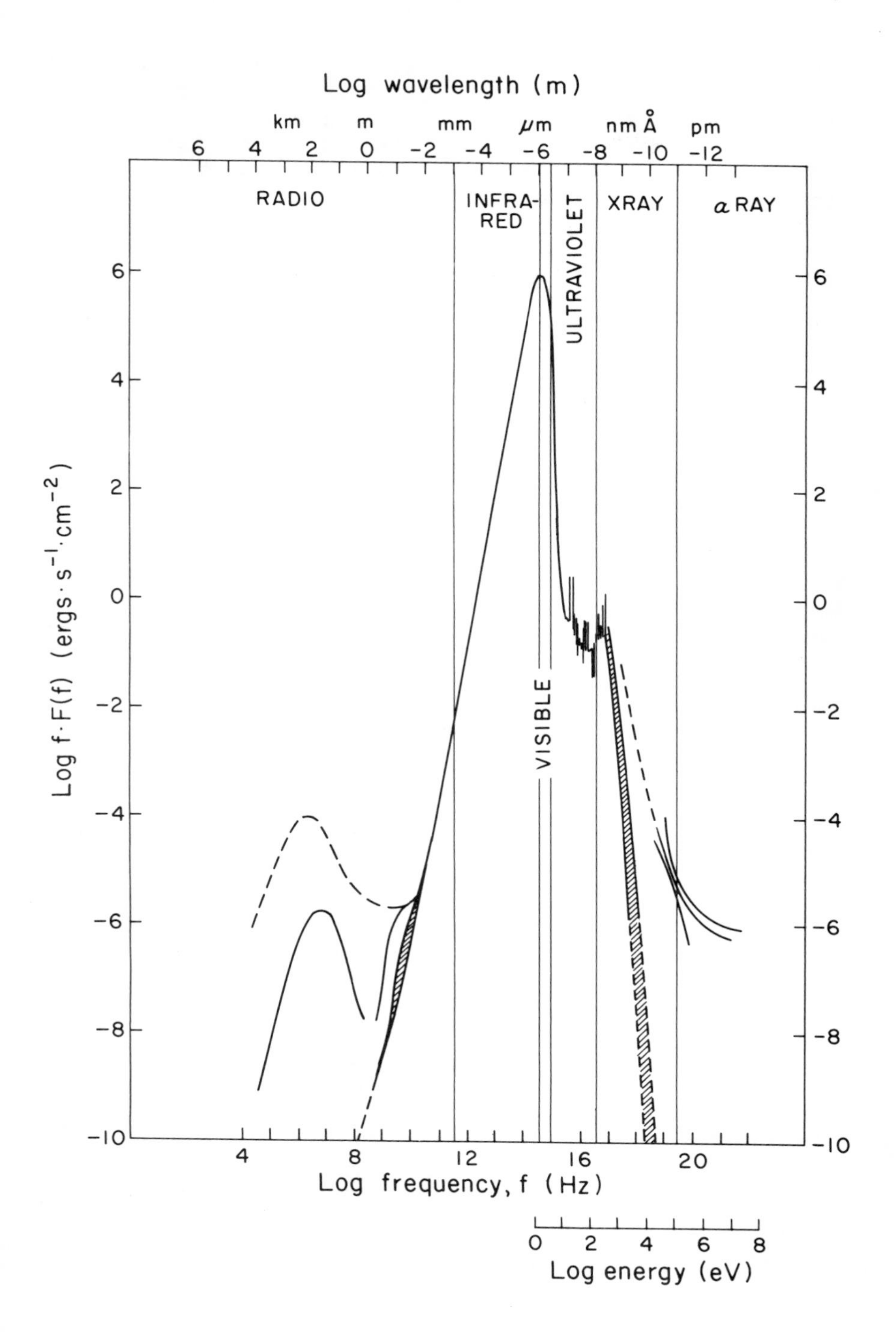

Log wavelength (m)
km
m
mm
μm
nm Å
pm
6
4
2
0
-2
-4
-6
-8
-10
-12
RADIO
INFRA-
RED
ULTRAVIOLET
XRAY
α RAY
VISIBLE
Log f·F(f) (ergs·s⁻¹·cm⁻²)
6
4
2
0
-2
-4
-6
-8
-10
4
8
12
16
20
Log frequency, f (Hz)
0
2
4
6
8
Log energy (eV)

the wide variation in radio fluxes is caused by variable distance of the source; at 30 kHz some sources envelop Earth and others retreat beyond the Sun. Spectra of brightness temperature (Steinberg et al. 1984) and of emissivity (Tokar and Gurnett 1980) of solar bursts are shaped much like the corresponding curves at the left of Figure 2.1. Solar-cycle and active-region variations are important in UV and EUV flux, whereas flare changes predominate in X rays and gamma rays (Fig. 2.2).

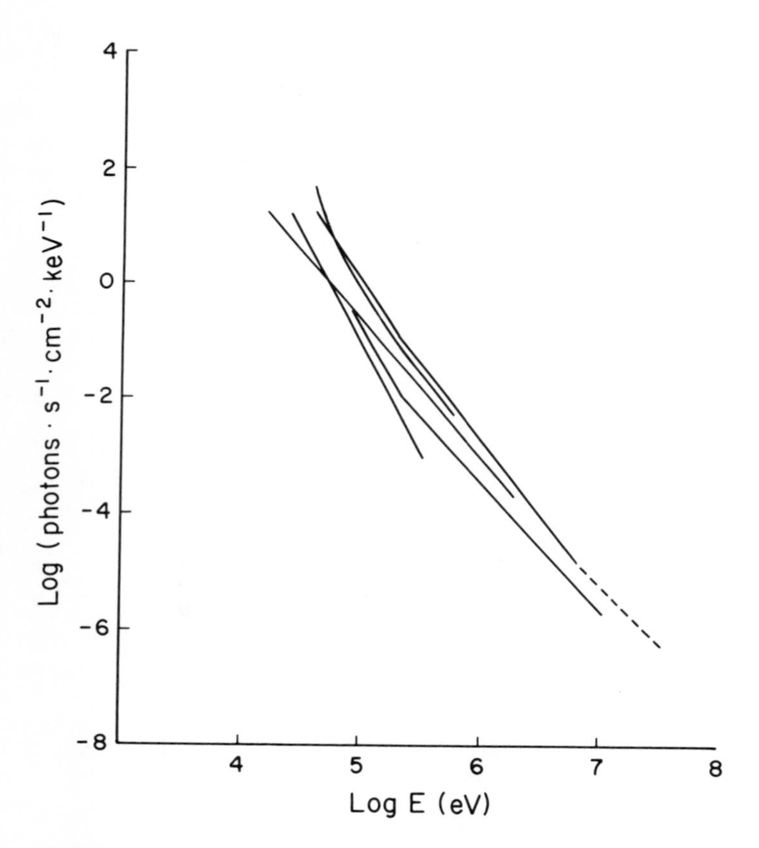

Figure 2.2--Energy dependence of flux of X rays and gamma rays for several flares. Data sources are Kane (1983), Nitta et al. (1983), Ohki et al. (1983), Rieger (1982), and Yoshimori et al. (1983).

A big flare in the chromosphere can cover a thousandth of the solar surface and remain bright for hours. "Microflares" have areas a hundred times smaller and last only a few minutes. Some flares, even from their remote location on the Sun, affect human activities (hence this report), but in the total flow of solar energy, integrated over the wavelength spectrum and over the solar disk, a flare is a minor fluctuation. According to studies of one typical flare by Canfield et al. (1980) and Webb et al. (1980), energy expended in ejecting mass exceeds radiated energy. The average mass flux due to flares may be a few percent of that in the ongoing solar wind. The rate of radiative energy from flares probably does not attain one part in ten thousand of ongoing solar radiative output.

The Active Cavity Radiometer Irradiance Monitor (ACRIM) on the Solar Maximum Mission measures total solar radiation. It shows fluctuations with amplitude 10 or 20 times larger than the above estimate, but these are week-long depressions associated with the presence of sunspots. No flare-related increases have been detected by ACRIM. Integrated flare energy cannot be more than a factor of 10 greater than the sum of measured or estimated energy already recognized in separate spectral regions (Hudson and Willson 1983). The new upper limit supports existing estimates and indicates that no large source of emission in the radiation spectrum has been missed.

Most efforts to forecast flares are based on the implicit assumption that the flare is a definable physical process or, if the process is partly physical and partly stochastic, that study of the physical part will be most rewarding. Rosner and Vaiana (1978) showed that attention to the stochastic part may also be useful, and that a statistic as simple as flare occurrence frequency as a function of flare energy can rule out certain modes of energy storage and release. They noted that the solar-flare occurrence rate follows a power-law distribution in energy and found that this is consistent with energy storage at a rate proportional to the stored energy but not consistent with energy storage at a constant rate. Their analysis and conclusions showed that something can be learned by compiling actuarial tables of solar-flare energy.

Energy must be estimated from available data, and these refer mostly to flare importance, roughly proportional to the logarithm of area, or of X-ray peak flux. From the imaged X-ray data analyzed by Pallavicini et al. (1977), one can deduce that a power law describes the occurrence rate of flares measured by their soft X-ray emission. In H-alpha, flare brightness and duration are both correlated with area in such a way that a measure of energy output, the product of brightness, area, and duration, is proportional to area squared (Sawyer 1967). Occurrence rate can be described as a power law in energy. The slope is such that flares within each H-alpha importance class contribute about the same amount of energy.

A power-law in flare energy provides a reasonably good fit to a recent small sample shown in Figure 2.3. For an increase of importance by one class, say 1B to 2B, area increases by a factor 2.5, so 0.4 is added to the logarithm of area, and 0.8 to the logarithm of energy. Number of flares decreases by a factor of 10, so log [N(I)/N(I')] ~ -1.0(I-I') ~ -1.2 (log [E(I)/E(I')]). At the right side of Figure 2.3 is a scale for counts of X-ray bursts in the same time interval. When activity is high, the total number of detected bursts above the lowest (C) level fails to increase as activity increases because small bursts do not rise above the enhanced background. The value N'(C) represents an attempt to compensate for this loss by extrapolating the distribution observed at a time of lower activity. The plot provides an estimate of H-alpha importance in terms of peak X-ray flux.

During active times, flares impose upon the flow of solar energy a sequence of minor fluctuations of varying amplitude and occurrence frequency. Of course, at the minimum of the activity cycle, these fluctuations disappear for months at a time. More needs to be known about the spectral characteristics of these fluctuations, how they vary with the activity cycle and from one active region to another, and how closely fluctuations in one part of the wavelength or energy spectrum follow those in another part. The biggest flares are known to cluster in location and time (see Section 3.7).

A physical model of solar flares is starting to emerge. Like the flare it attempts to describe, the model will certainly be formidably complex. In its most simplified form, a flare model must describe how a nonuniform atmosphere responds hydrodynamically and thermodynamically to energy input. Excitation of electrons in atoms and ionization must be calculated, and emission and absorption of radiation described. Motion of energetic charged particles and their exchange of energy with the ambient plasma must be considered. Relevant length scales cover a range of 10^{18} or more, from the diameter of an electron to that of the flare itself (which represents an even larger scale because it sends mass and radiation beyond the orbit of Earth).

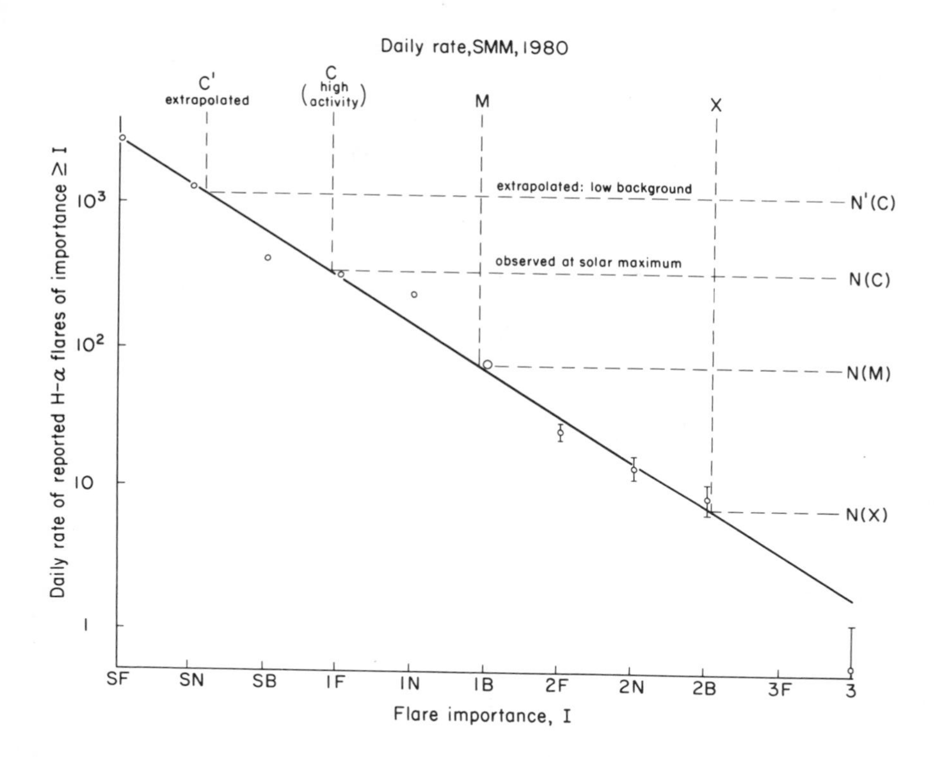

Figure 2.3 The daily rate of flare reports at the recent activity maximum, March through September, 1980, as a function of flare importance. For each H-alpha importance class, the number of reports in Solar-Geophysical Data of flares of that class or greater is divided by the number of days, 184, in the sample. The size of the circle or the error bar represents the expected variance (the square root of the counted number). Counts of X-ray bursts with peak flux equal to or greater than each limit for the same period are shown at the right. (X-ray data courtesy of K. Kildahl, SESC, NOAA.)

The solar atmosphere, even undisturbed, is sufficiently complex that physical description is still incomplete. Available models are empirical--they translate observations of radiation intensity and spectral quality into the physical quantities density and temperature. In Figure 2.4, temperature is plotted as a function of density. Significant features are (1) the temperature minimum, (2) a temperature plateau in the lower chromosphere where hydrogen

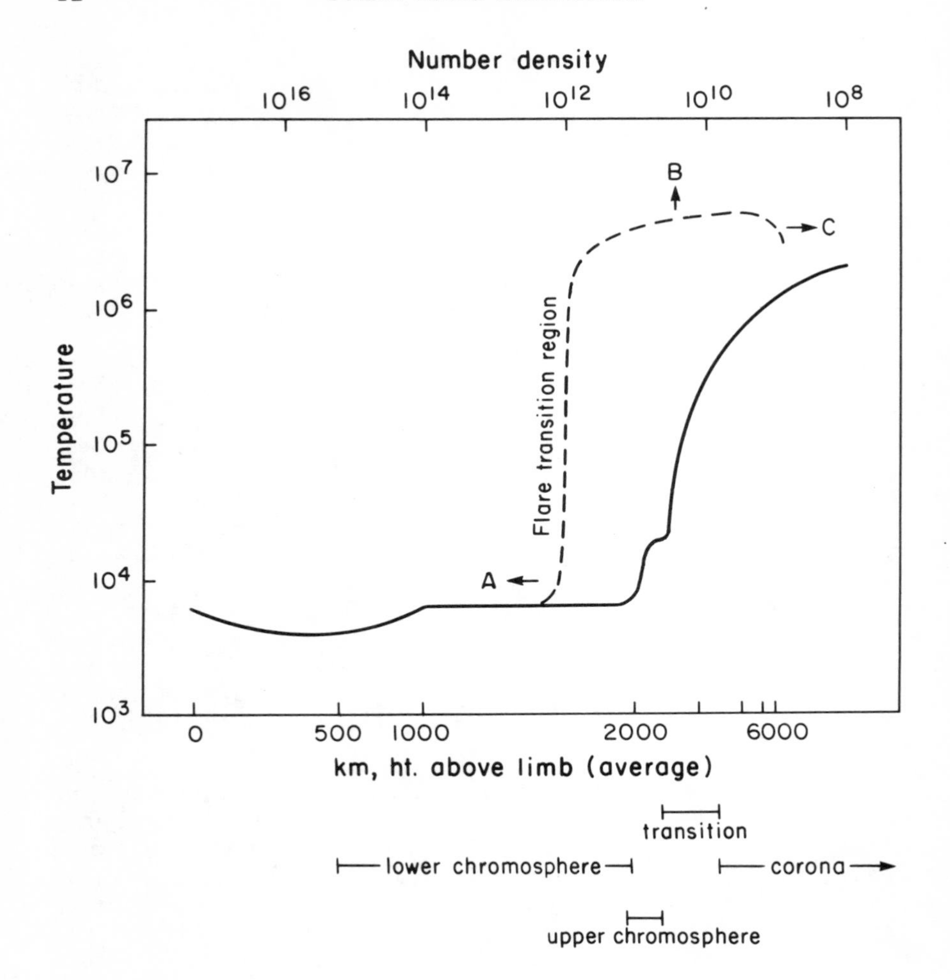

Figure 2.4--Temperature profiles in models of the undisturbed solar atmosphere (solid curve) and of a flare (dashed curve). A scale of number density (atoms or ions) appears at the top, and a scale of typical heights at the bottom. Features of hydrodynamic models of flares are (A) downward moving compressive wave, (B) heating of the flare plasma, and (C) upward expansion of heated plasma ("evaporation") and associated upward moving shock.

ionization consumes the available energy, and (3) a second temperature plateau in the upper chromosphere where energy loss through radiation in the Lyman-alpha spectral line becomes effective. The transition region is a remarkably thin layer between the upper chromosphere and the million-degree corona. A scale of heights at the bottom of Figure 2.4 reflects the exponential decrease in density with height. The scale of height can be very different from one place (and time) in the chromosphere to another. Consider, for example, that a spicule can raise the transition region as much as 20,000 km above the average height, pushing it far beyond the limits of Figure 2.4. This topological complexity means that for a model, a scale of density or of optical depth is more practical than a scale of geometric height.

In their flare model, MacNeice et al. (1984) described initial conditions with a model chromosphere and transition region like that in Figure 2.4 but Somov and coworkers (1981, 1982) started with a simpler, isothermal initial atmosphere and arrived at a similar flare model. Each envisages an elementary tube of magnetic flux, with scale about 10,000 km in height and 1000 km in cross section. Somov's flux tube is vertical, whereas MacNeice's is semi-circular, with two feet in the chromosphere. In all these models, the role of the magnetic field is simply to contain the plasma; the magnetic field is not included in the equations that describe energy balance or motion. Therefore, these models describe the hydrodyamics and thermodynamics of a compact, contained, "low-beta" flare.

Somov et al. (1982) postulated a thermal energy source, simply described by letting the temperature at the top of the chromosphere rise to 100 million degrees and decay over a 10-second time interval. The other two models specify intensity and energy spectrum of a beam of energetic electrons injected during a 10-second time interval at the top of the flux tube, then calculate the transfer of energy from these downward streaming electrons to the ambient plasma. Although energy is transferred by a modified form of thermal conduction in one case and by charged-particle encounters in the other, the results have many similarities.

The models predict formation of a flare transition region where temperature jumps from thousands to millions of degrees. Below the flare transition region, plasma is sufficiently dense that added energy is promptly radiated away. Above the flare transition region, plasma of low density radiates inefficiently. When energy is added, temperature rises abruptly, pressure increases, and the heated plasma expands and pushes upward within the flux tube into the corona (see Figure 2.4).

Below the flare transition region, a heat front propagates downward. Within seconds, its speed exceeds the ambient ion-acoustic speed. A shock forms and propagates into the chromosphere, compressing the chromospheric plasma. Radiative cooling is enhanced, and a cold condensation moves downward as a supersonic compression wave.

Special physical processes occur at a steep gradient of temperature. Heat is transferred less efficiently by classical conduction than by energetic electrons that move between the two regimes of temperature (Gurevich and Istomin 1979). Heat flux is limited by possible instability, and electron-ion coupling becomes important. Effects of two separate temperatures may be observed (Smith and Auer 1980). Without dwelling on the difficulties of this work--questions of physical and computational instabilities, and the fine grids required for numerical resolution of impulsive changes and sharp spatial gradients--we note that each of the hydrodynamic simulations predicts these main features: a flare transition region at the density critical for radiation loss, a compression wave moving downward into the chromosphere, and a piston-like upward expansion of heated chromospheric plasma into the corona.

On the basis of the temperature, density, and velocity found from the hydrodynamic analysis, one can calculate ionization of each chemical element and population of atomic levels. Then it is possible to predict flare radiation of various types from each location at each time. Simplifying assumptions are clearly necessary. For example, dynamic analysis by MacNeice et al. (1984) shows that a steady-state analysis would be adequate in many circumstances. In theory, spectra of X-ray bursts might provide a diagnostic to distinguish the thermal model from the electron-beam model because an isothermal source yields an exponential spectrum, and energetic electrons yield a power-law spectrum. In fact, interpretation is not so simple. Observed spectra often can be described as a combination of power laws with different slopes, and this could indicate either a multithermal source, or a complex process of acceleration.

The rest of this chapter describes observations and physical descriptions of individual flares. The remarkable rate of progress in physical understanding might lead to the expectation of similar progress in ability to forecast flares, but this expectation has not been realized. This calls for examination of observational and theoretical advances and their possible relevance to prediction.

After this examination, an interpretation of the seeming paradox is offered: although concepts of the flare process have been extended and clarified, even revolutionized in some areas, most of these concepts are still qualitative. Practical prediction demands accurate quantitative definition of relations among physical quantities, which themselves must be well defined. Flare physics is still in the stage of identifying significant problems and parameters; success in measuring and relating these quantities lies mostly in the future. Understanding is growing fast, but still is far from the completeness needed to aid prediction.

Although a complete theory is still out of reach, recent advances provide a coherent framework for planning further effort toward that goal. A series of steps in the flare process can be discerned and outlined as follows.

Large-scale circulation organizes velocity fields and concentrates magnetic flux in temporal-spatial wave-like patterns that include the 11-year activity cycle.

Active regions are concentrations of magnetic flux. Non-potential magnetic-field configurations contain excess energy that can be released in a transition to a state closer to the minimum-energy potential field. Emergence of additional magnetic flux is likely to increase complexity and available energy, and flux disappearance can lead to instability. Velocity shear near or below the photosphere, where the magnetic field is strong but controlled by even stronger forces of the moving plasma, produces magnetic shear that further stresses the field and adds to the non-potential component and to available magnetic energy.

Instability can arise from a small perturbation and lead to field-line reconnection and rapid relaxation of the stressed magnetic field with release of part or most of the available energy. Initially, excess magnetic energy is transformed rapidly within a small volume into kinetic energy of charged particles.

The **impulsive phase** marks the transfer of this kinetic energy to the ambient atmosphere where ionization, excitation, and heating result in bursts of radiation that occur throughout the spectrum, especially strongly in the soft X-ray and EUV ranges. Electrons spiraling in magnetic fields emit microwave radio bursts. Collisions of accelerated particles with ambient ions, electrons, and neutrons produce X rays and gamma rays.

Electrons that escape into the high corona and solar wind induce plasma oscillations and bursts of radio emission at metric and longer wavelengths.

The **late gradual phase** starts when ambient gas, heated by interaction with energetic electrons and protons and initially confined within a magnetic loop, expands into the corona. Continued particle acceleration and interaction, now with this compressed hot coronal plasma, yields a new broad spectrum of radiation that differs from that of the impulsive phase, depending on the energy and direction of motion of the fast particles and on properties of the ambient plasma: density, temperature, volume, motion, and magnetic field.

Coronal mass ejection accompanies most energetic flares and accounts for much of the energy released. Over the active region, outward streaming plasma combs magnetic field lines into an empty, smooth, nearly radial configuration after the event. Associated with mass ejection are prominence eruptions, type II radio spectral bursts, long duration X-ray enhancements and coronal and interplanetary shocks.

Second-stage acceleration of energetic charged particles in some flares may indicate renewed acceleration, probably in shock waves and possibly related to reconnection of magnetic field lines torn apart by mass ejection or by expansion of hot plasma in the gradual phase.

In the next sections some recent advances in flare physics are described in the framework of the flare process outlined above. Although the path is not yet obvious, advances such as these should lead eventually to the quantitative understanding of the flare process that is needed for forecasting flares and their effects on Earth and in space.

2.2 LONG-LIVED, LARGE-SCALE MAGNETIC AND VELOCITY FIELDS

Understanding flares and understanding all solar activity begins with the sunspot cycle (Figs. 2.5 and 2.6). Recent observations and analysis link the formation of active regions to large-scale patterns of magnetic fields (Fig. 2.7) and velocity fields on the solar surface and evidently to a fundamental internal oscillation. Recurrence

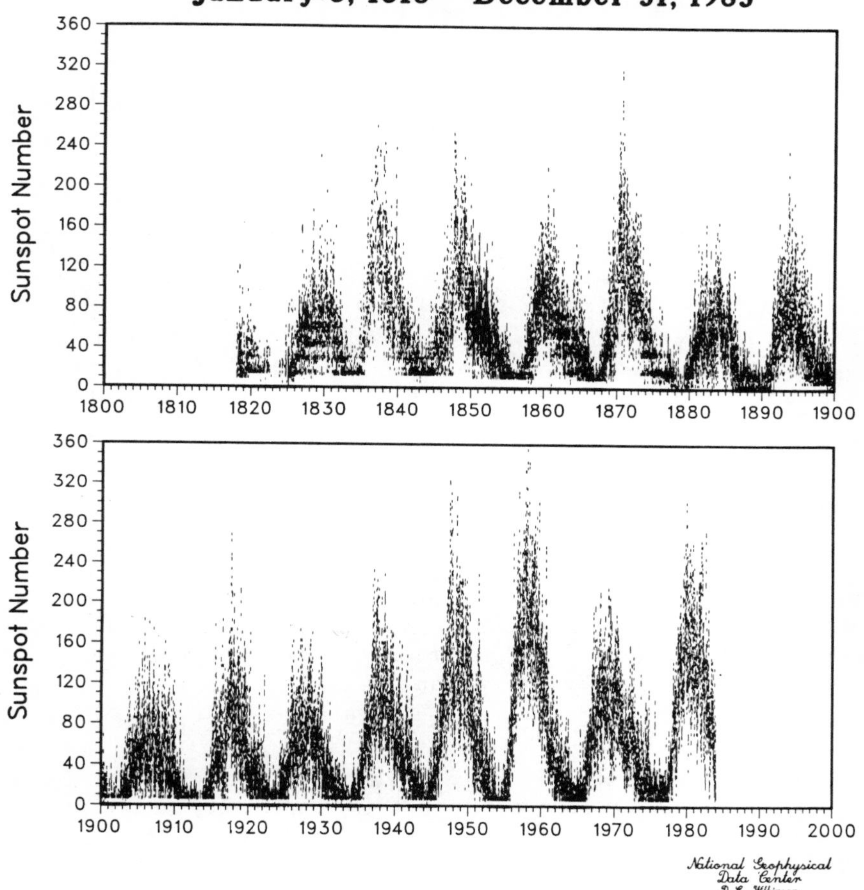

Figure 2.5--Plot of daily sunspot number, displaying the magnitude of day-to-day variations, as well as the 11-year periodic variation. Sunspot number is defined as the weighted average of the estimates by different observers of the quantity: (number of sunspots + 10 x number of sunspot groups).

of active regions at preferred heliolongitudes, sometimes spanning more than one 11-year cycle, has been noted over the years; early work on "active longitudes" was reviewed by Sawyer (1968). Howard and Bumba (1965) documented evidence

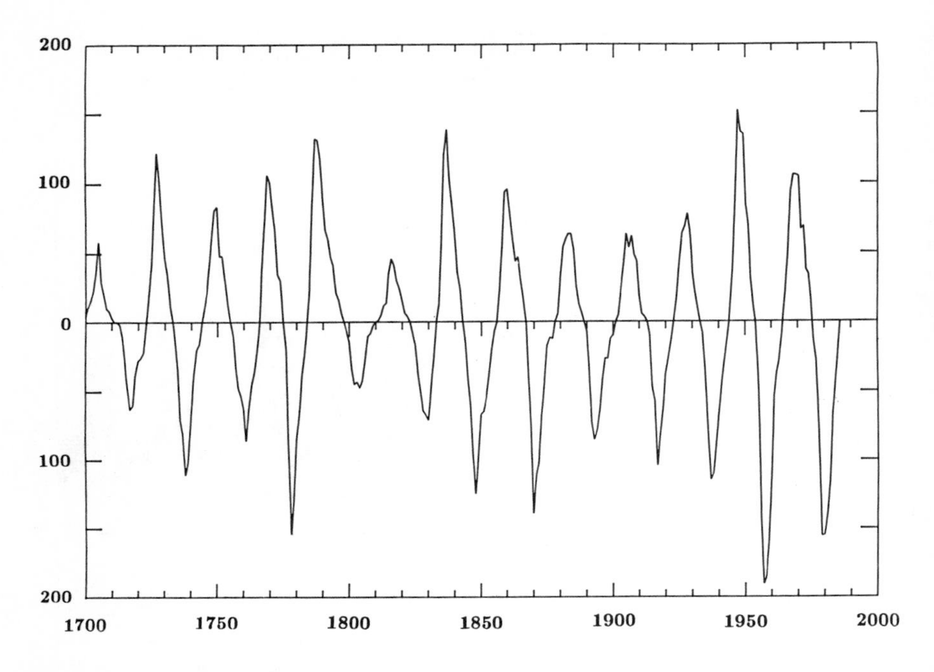

Figure 2.6--In this display of the solar activity cycle, alternate cycles take opposite signs. This is one of three modifications found by R.N. Bracewell (1985) to enhance the significance of mathematical analysis.

for large-scale, long-lived complexes of the magnetic field itself. Recent studies confirm this continuity and indicate a connection existing over years, and even over 11-year cycles (Martin and Harvey 1979, Legrand and Simon 1981, McIntosh 1981, Gaizauskas et al. 1983, Leroy and Noens 1983, Stewart 1985).

Leighton (1959) modeled dispersion and migration of active-region fields; Babcock's (1961) model builds magnetic fields of the new cycle from the migrating fragments of old-cycle fields. Is the observed persistence of activity more than the result of surface fields participating in large-scale random walks and chance clusterings? Gaizauskas et al. (1983) found that magnetic flux persistently emerges in the active region and disappears in the active region; only a small fraction of the flux migrates or disperses, as envisaged in the Leighton model. Dicke (1978) searched long-term records for signs of periodicity and concluded that indirect evidence supports the presence of an internal

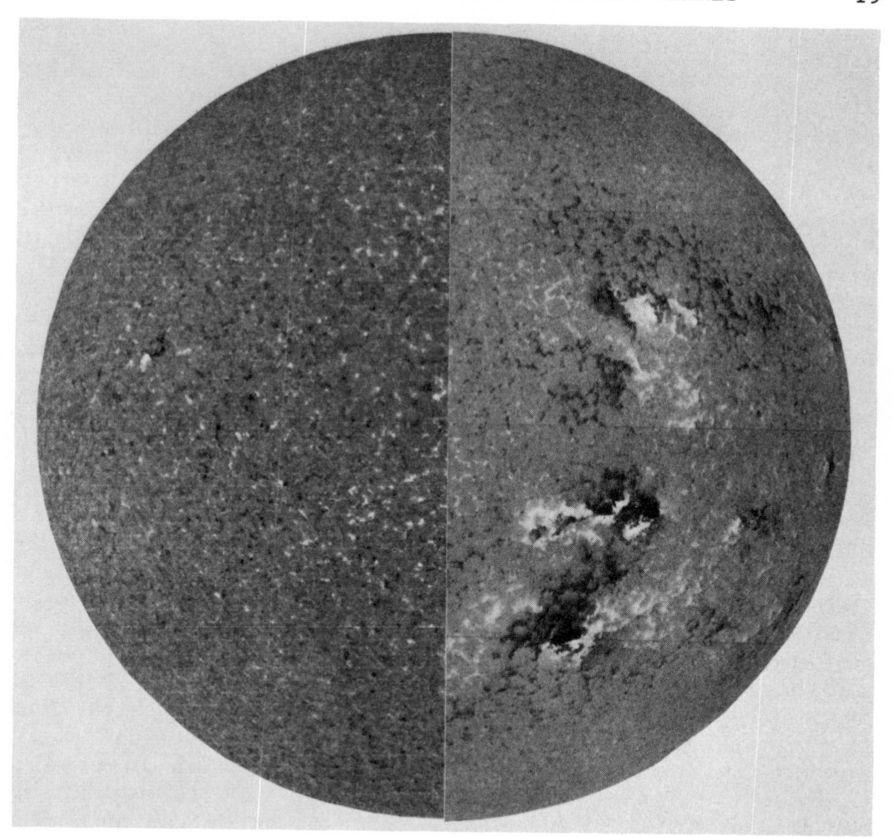

Figure 2.7--This collage shows magnetic fields on the surface of the quiet and active Sun. The left-hand image, made near cycle minimum, is devoid of active regions. The right-hand image, made near maximum, shows active regions north (top) and south of the equator. One can see a general tendency for black (-, toward Sun) polarity to lead in southern active regions and white (+, away from Sun) polarity to lead in northern regions, but with considerable complexity, especially in the southern activity complex. (Courtesy of Jack Harvey, National Solar Observatory.)

solar chronometer that regulates the phase, as well as the period, of the growth and decay of magnetic fields on the Sun's surface.

Large-scale, long-lived organization of both velocity fields and magnetic fields appears in recent results of statistical analyses of long series of observations (Figure

2.8). Superposed on the average (rigid) rotation are faster and slower currents. Starting at a high latitude near the time of maximum activity as a "polar spin-up," a fast current in each hemisphere moves equatorward with time, dying out at a low latitude near the time of the next activity maximum. These propagating waves of enhanced rotational velocity can be described as a torsional oscillation: at a given latitude, rotational speed varies with time from faster than average to slower than average--or differential speed swings from west to east--with a period of 11 years. Each stream drifts from pole to equator in the same period. Magnetic flux emerges and sunspots appear at the shear zone poleward of the fast flow, and this preferred zone for sunspot birth moves equatorward in the course of the sunspot cycle, forming the familiar "butterfly diagram" of activity migrating to lower latitudes (LaBonte and Howard 1982, Howard and LaBonte 1983, Snodgrass and Howard 1985). The details of the pattern revealed by subtracting average rotation depend on whether rigid rotation (the pattern described above) or latitude-dependent differential rotation is subtracted. Either analysis describes the complex evolution of magnetic and velocity fields in terms of a much simpler wave motion. In the discussion following Howard's 1983 I.A.U. Symposium presentation, Giovanelli remarked: "I think that the torsional oscillation is the most important discovery concerning the solar cycle that has been made in recent times. It contains the clue to the whole cycle mechanism. Our thanks should be due to the whole Mount Wilson team for this significant new phenomenon."

What drives the torsional oscillation? Now there are only faint clues that may lead to tomorrow's answers. Filtergrams of successively weaker absorption lines reveal successively deeper layers of the atmosphere. They provide direct images of atmospheric structure at different heights.

Figure 2.8 (facing page)--Large-scale, long-lived velocity and magnetic fields. Synoptic charts of (a) the full torsional oscillation pattern, determined by subtracting the overall average of the absolute differential rotation from the differential rotation curve determined from the corrected slopes, and (b) the total (positive plus negative) magnetic flux. All curves represent 20 Carrington rotation averages. The contour levels in (a) are 2, 4, 6, 8, and 10 m/sec, and in (b) they are 0.5, 1, 2, . . ., 9 x 10^{20} Mx. (From Snodgrass and Howard (1985) by permission of Science magazine. Copyright 1985 by the AAAS.)

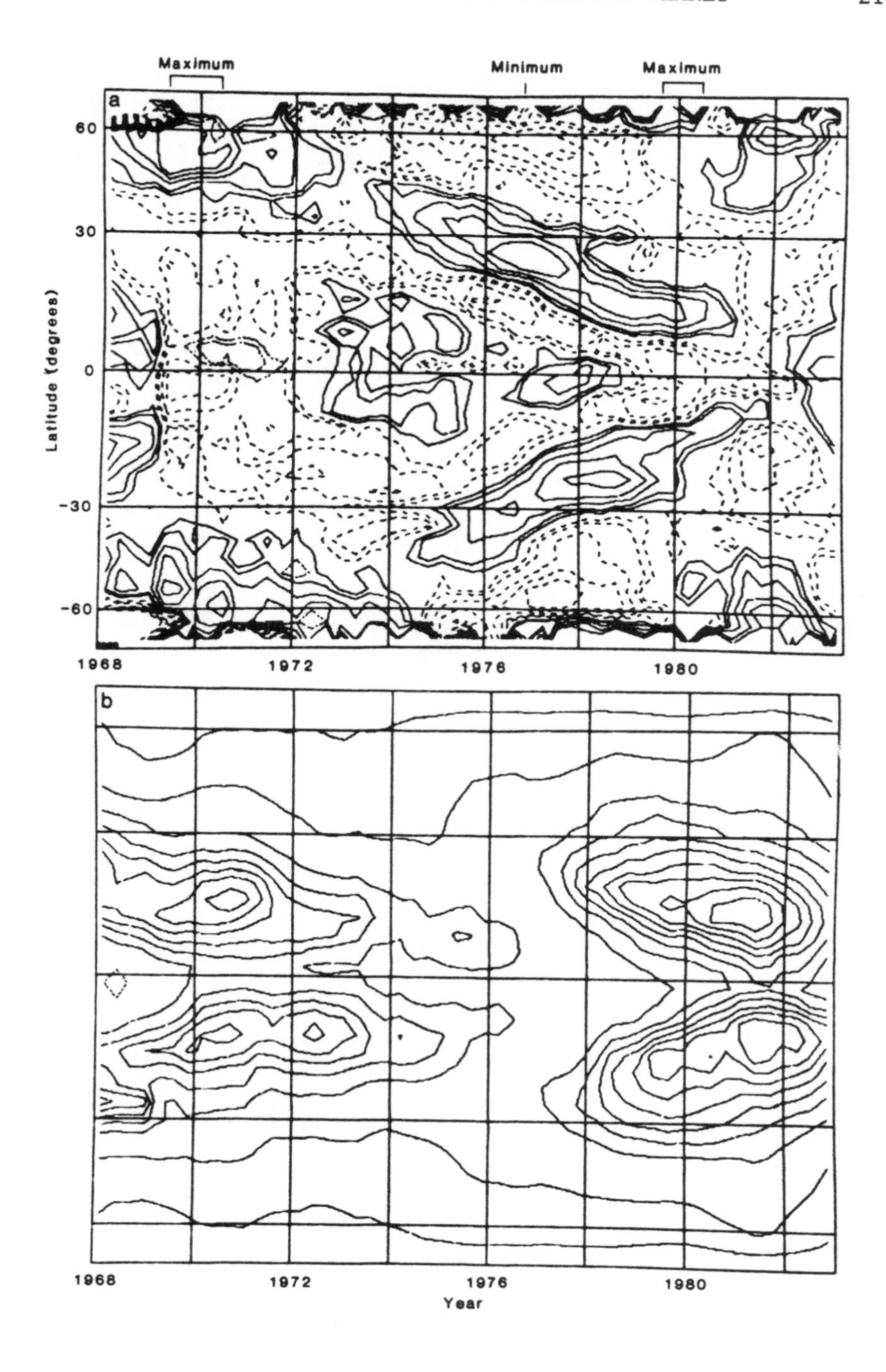
Maximum
Minimum
Maximum
a
60
30
0
-30
-60
Latitude (degrees)
1968
1972
1976
1980
b
1968
1972
1976
1980
Year

The lowest imaged layer is at the bottom of a cool sunspot umbra; the layer below is opaque to all radiation except the agile neutrino. The solar interior is not a complete mystery, however. Internal conditions and structure are deduced from basic physical theory such as conservation of mass, momentum, and energy; from the flux of escaping neutrinos; and from properties of waves that propagate through the interior to produce observable oscillations at the surface. The central core contains most of the mass of the Sun. In this hot, dense environment, nuclear fusion occurs. In each set of reactions the input is 4 protons and 2 electrons, and the net output is an ion of helium, some 20 to 30 MeV of radiated energy in the form of gamma rays, and 2 neutrinos that promptly carry away about 1 MeV of energy, but no charge or mass. Each electron pairs with a positron that is produced when 2 protons combine. Thus the output has less mass and more energy than that put in. In the middle layer, the radiative interior, the radiated energy is repeatedly absorbed and reemitted, with net outward flux and net decrease of average photon energy and frequency. Energy is transported by convection to the photosphere, the thin skin within which the Sun becomes opaque.

In the chromosphere, temperature rises again. Temperature plateaus in Figure 2.4 mark the level (1) where the ionization of hydrogen atoms consumes the available energy, and (2) where temperature first becomes high enough to permit radiation in the important spectral line, Lyman alpha. At wavelengths increasingly close to the center of increasingly strong absorption lines, chromospheric structure can be imaged at higher and higher altitudes up to 0.03 solar radius. Structure in the hot, ionized corona is revealed in emission lines, in electron-scattered photospheric light, in radio emission from electrons excited in various ways, and by active radar.

Oscillations of brightness and velocity at the solar surface have periods from minutes to hours. Measurement, analysis, and interpretation of these oscillations is called helioseismology because it reveals the structure of the solar interior much as seismology reveals the structure of Earth's interior. Helioseismology has already provided important information; it has confirmed the density and temperature structure deduced from physical models, determined the depth of the convection zone, deduced a value of the initial abundance of helium, and provided information about mixing of material in the core. Outstanding questions of special interest to understanding solar activity concern the rotation rate of the core, variability of the rate and its phase stability, a possible strong magnetic field in the core, and velocity shear produced by variation of the rotation rate with latitude and with depth. Advances in helioseismology require long uninterrupted series of observations, which may be acquired in the future through a worldwide observational network and from a continuously sunlit satellite (Noyes and Rhodes 1984).

2.3 MAGNETIC-ENERGY BUILDUP IN AN ACTIVE REGION

Magnetic-energy buildup is the part of the flare process of obvious relevance to flare forecasting. Because recent observational advances are described in Chapter 3, on flare precursors, emphasis here will be on theory and interpretation.

Magnetic-field reconnection as the primary energy release process in solar flares has gained credibility through its direct observation in Earth's magnetosphere (Hones 1984; Sonnerup et al. 1984). The Hones volume on reconnection in the magnetosphere is dedicated to the solar physicist Giovanelli, as a tribute to his early recognition of the significance of magnetic reconnection in solar flares. Now flare theorists are reaping benefits from in situ observations of reconnection in the magnetosphere. Figure 2.9 illustrates a simple case of magnetic reconnection.

Geomagnetospheric reconnection has some aspects relevant to solar flares: reconnection requires an electric field along a boundary between distinct magnetic regions or cells. Plasma and magnetic flux transfer from the merging cells to the separating cells (Figure 2.9a). In "driven" reconnection, plasma is continually supplied to the merging cells and removed from the separating cells by an external driver--the solar wind for the magnetosphere.

Another flare-relevant observation is that large plasmoids containing closed magnetic loops move rapidly out the magnetotail at the time of geomagnetic substorms. They are identified as magnetic bubbles that are closed and then released by reconnection. Seemingly analagous magnetic bubbles in interplanetary space were identified by Burlaga and Klein (1982), who pointed out similarities to coronal mass ejections associated with some solar flares. Figure 2.10 compares the formation of a plasmoid in the magnetotail to a familiar situation. Surface tension separates the water drop; tension in magnetic lines of force is a factor in plasmoid formation. Magnetospheric substorms were compared quantitatively to solar flares by Haurwitz (1972). Figure

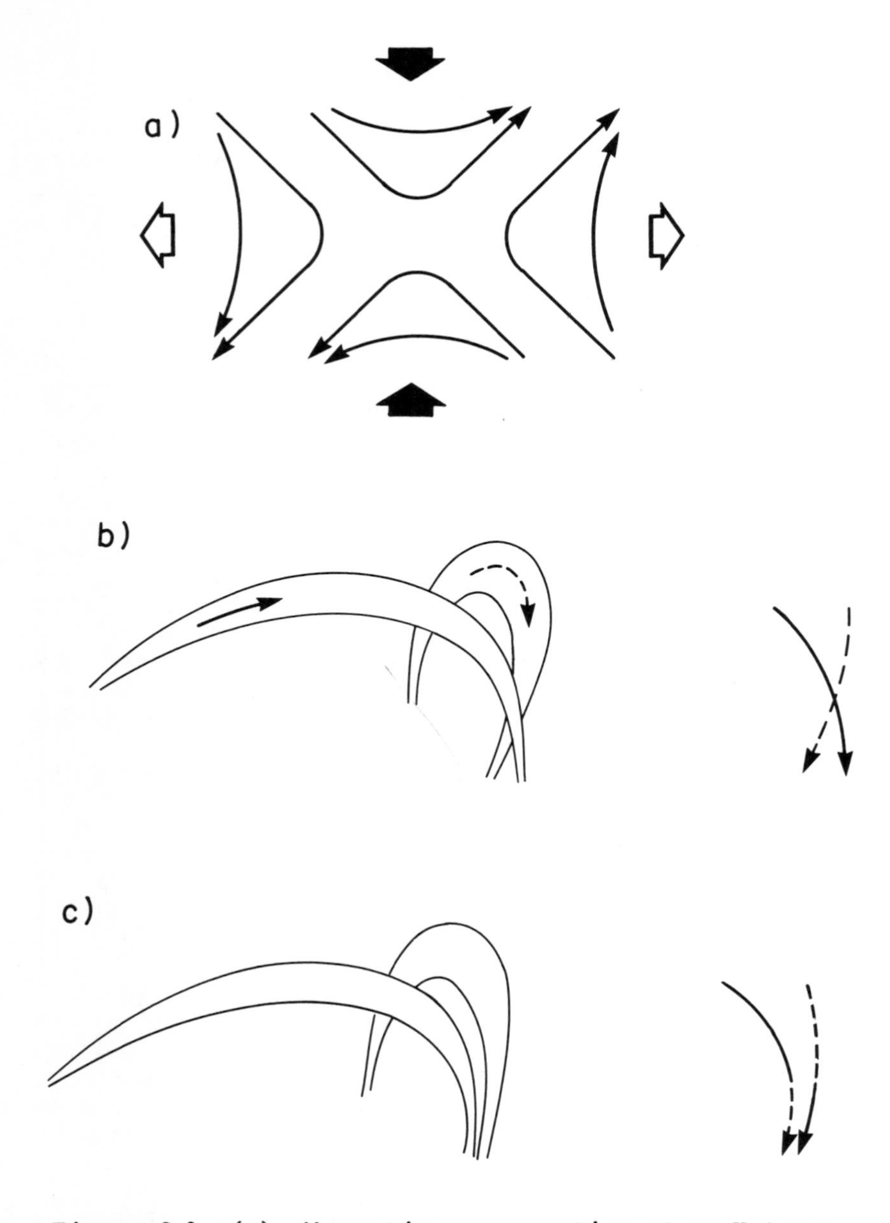

Figure 2.9--(a) Magnetic reconnection at an X-type neutral point. The curves represent magnetic lines of force and the heavy arrows represent flow. (b) A possible analogous arrangement of magnetic flux tubes at the Sun's surface. Arrows indicate the direction of the magnetic field. (c) The result of reconnection.

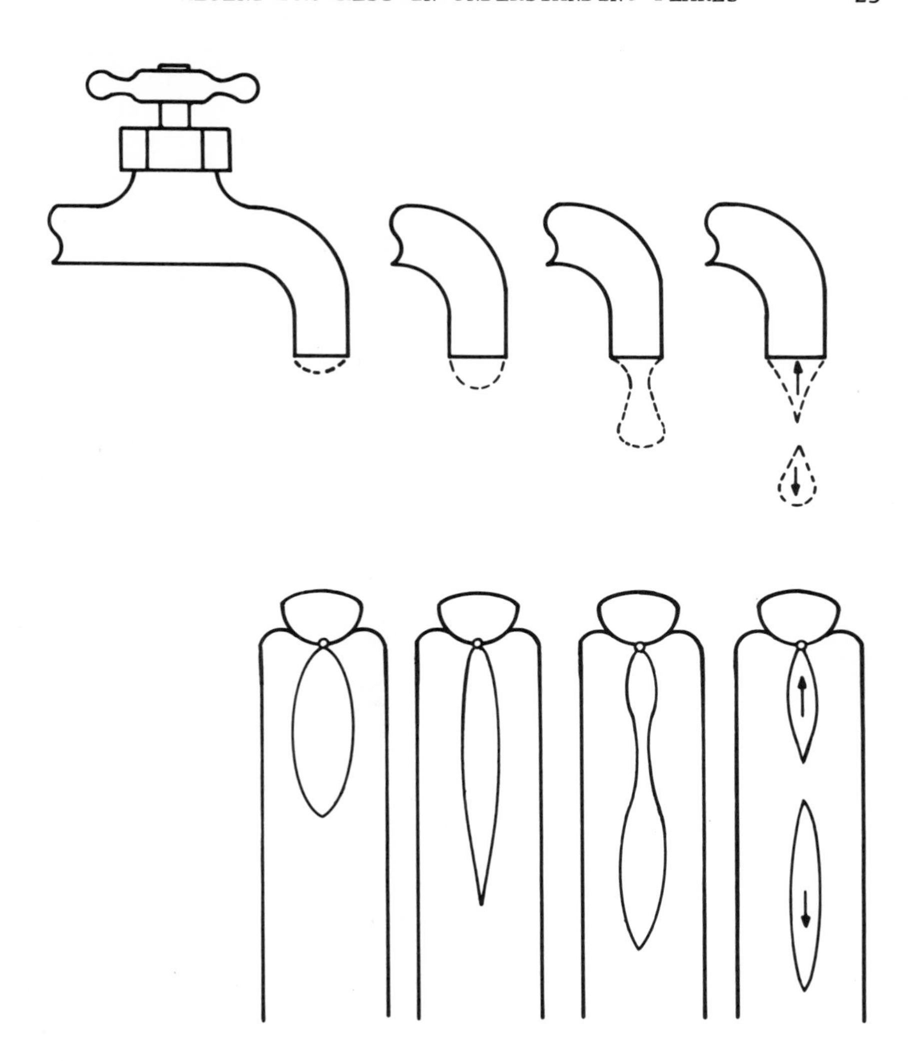

Figure 2.10--Analogy of plasma-sheet severance and plasmoid formation to the behavior of water dripping from a leaky faucet. (Reproduced from Hones (1979) with permission of the American Geophysical Union.)

2.11 illustrates solar-magnetospheric similarities noted by Priest (1984). Magnetic reconnection in the dayside magnetosphere is compared to reconnection of emerging magnetic flux at the solar surface, and reconnection at an X-line--a continuum of X-points--explains both an erupting prominence

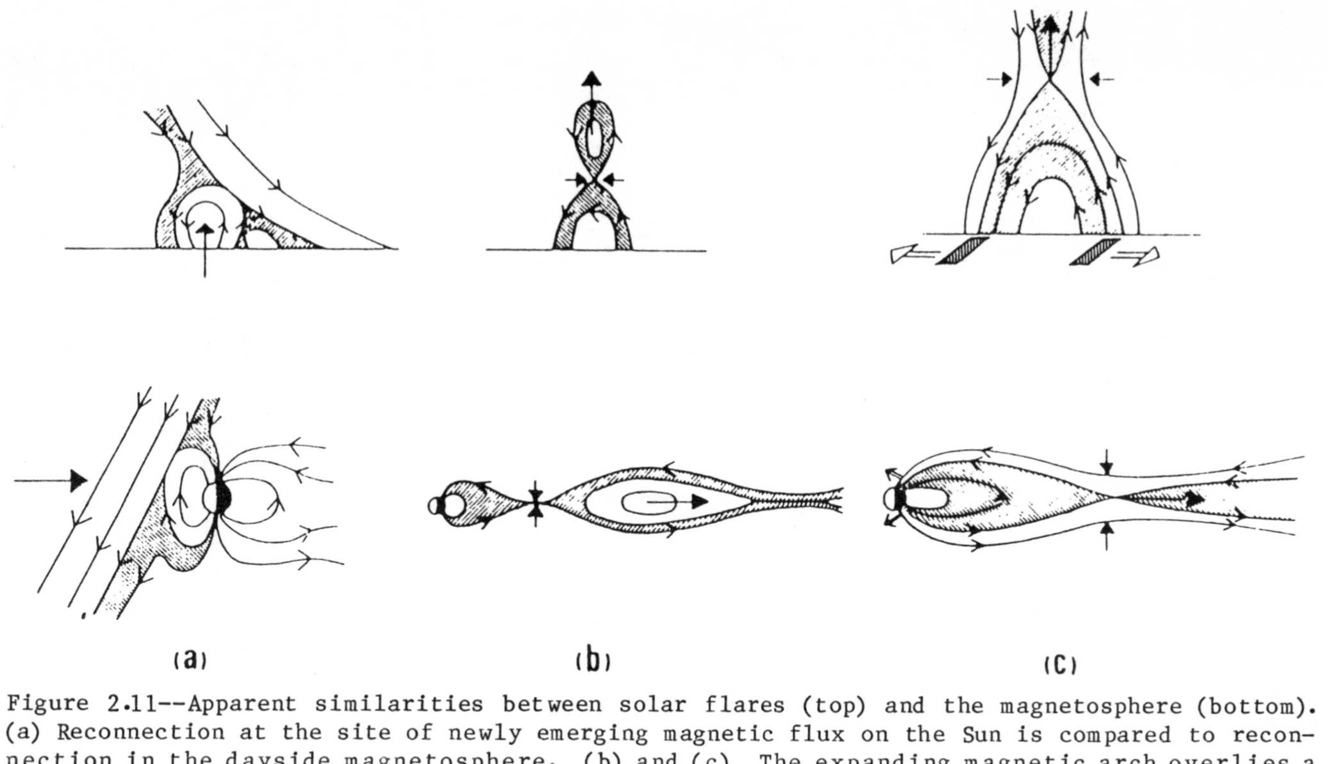

Figure 2.11--Apparent similarities between solar flares (top) and the magnetosphere (bottom). (a) Reconnection at the site of newly emerging magnetic flux on the Sun is compared to reconnection in the dayside magnetosphere. (b) and (c) The expanding magnetic arch overlies a two-ribbon flare represented by parallel surface brightenings, moving apart in (c). A rising prominence is pinched off at the Sun, an escaping plasmoid in the magnetospheric substorm. (Reproduced from Priest (1984) with permission of the American Geophysical Union.)

on the Sun and plasmoid escape in a magnetospheric substorm. Priest points out that in flares and substorms different physical effects are present as well as important differences in geometry and in values of physical parameters. Nevertheless, some processes believed to be important in flares are observed in magnetospheric reconnection: particle acceleration, plasma waves, and evolution of magnetic-field structure.

A lively controversy about the interpretation of magnetospheric substorms is significant also for solar flares. Is energy stored at a slow rate over a long period of time, then suddenly released, or does the rate of release respond to a highly variable rate of input? In the latter case, input rate should be the main determinant of output rate, but measuring rates and establishing a correlation has proved difficult even for the relatively accessible magnetosphere (Akasofu 1985).

Flare Theory

Is input rate always much slower than output rate during a flare, so that the total amount of energy released in the event is very much greater than the amount that could accumulate during the event? Spicer and Brown (1981) recognized this as a basic question, citing Sturrock's (1966) argument for the necessity of accumulation, but stating that the historical belief in energy storage and triggered release has "no observational justification at the present time."

Spicer and Brown (1981) organized their review of flare theory around physical mechanisms and their energy sources. Energy storage in magnetic fields implies the flow of electric current. Because current directed across the field exerts a force, any current must flow parallel to the force-free field. A magnetic field that is nearly force-free and scarcely influenced by the plasma it contains must have much greater energy density than the plasma: the ratio, beta, of thermal to magnetic energy density is low. A nearly force-free field will be closed, forming a loop or an arcade. Low-beta flare mechanisms release a small fraction of total available magnetic energy; they involve impulsive heating and continued containment of the plasma by the field. These are the conditions for a compact flare that remains within the confines of closed magnetic-field structures in the low corona above an active region, the situation that has been modeled in most detail. Perpendicular currents, high beta, and open field lines characterize fields that are not force-free, but are shaped by the motion of the plasma. This is the situation in flares that include coronal mass ejection, and perhaps in large, high, "two-ribbon" flares in general. It also characterizes conditions in a magnetic neutral sheet, as in a coronal streamer.

Of the flare-initiating mechanisms examined by Spicer and Brown only tearing instability can operate with a weak driver, with energy storage at the flare site. Each of the other mechanisms requires strongly driven currents and a continuing source of energy. This theoretical insight contrasts with the appealing, widely held, and often expressed view that energy is stored in the coronal magnetic field over a period of hours or days and then suddenly released at flare time by some "trigger," possibly spreading to a larger volume through a domino-like effect. Of course, proponents of triggered release can interpret this as evidence that the tearing instability is the operative flare mechanism. If gradual storage is the case, gradual expansion of the magnetic container, rather than catastrophic rupture, might be expected. Spicer and Brown asked--as did Warwick in 1962--"what could be more reasonable than to attribute photospheric or subphotospheric energy sources as the primary cause of flares in view of the fact that the photosphere is 'wagging' the corona and not the inverse?" Indeed, the venerable current-interruption model (Alfven and Carlqvist 1967) invokes strongly driven currents in the photosphere or convection zone.

A concept of significance to observation and theory is that of an elementary magnetic flux tube. Wang et al. (1984), at Big Bear Observatory, followed disappearing elements and determined the smallest observable flux. Typical magnetic elements, followed in a movie that shows their interactions, have flux more than 100 times greater than this observational limit. For one of these typical elements the flux is $10^{18.4}$ maxwells (Mx = gauss x cm^2) (Zirin 1985, Martin et al. 1985, Livi et al. 1985), a result expected from earlier work (Livingston and Harvey 1969). Comparison of the observations of magnetic elements with a simultaneous H-alpha movie showed a striking example of a small flare coinciding in time and location with the cancellation or submergence of a pair of flux elements. The sequence strongly suggests that H-alpha emission follows transfer of energy to the ambient plasma from the simplifying magnetic field. Flux cancellation is discussed further in Section 3.7.

Rust searched for evidence of magnetic-field change in the photosphere (Rust 1976) and in the chromosphere (Rust 1984) following large flares and concluded that "the net effect of flare energy release . . . is below the sensitivity of present methods of detection." Although postflare changes were found in many cases, similar changes occurred almost as frequently in the absence of flares.

Thus neither observation nor theory requires local accumulation of energy and triggered release. Recent progress includes description of basic mechanisms with necessary conditions as well as implications for processes such as particle acceleration. Much of this progress carries over to improved understanding of the flare's impulsive phase.

2.4 FLARE INITIATION: IMPULSIVE AND LATER PHASES

More Theory

Spicer and Brown (1981) gave some general requirements for flare initiation. Because microinstabilities saturate within times much shorter--and total energies much less--than in a typical flare, Spicer and Brown concluded that the instability must be driven throughout the lifetime of the flare by an external energy source. They noted that dense, strongly driven current implies an electric field strong enough to accelerate charged particles.

An illuminating discussion of new observations and theory is provided in the report of the Flare Precursor and Onset study group of the Solar Maximum Mission (SMM) workshop. Van Hoven and Hurford (1984) described a "theoretical preflare scenario" that emphasizes the significance of magnetic-field orientation relative to gradients of temperature and velocity in the plasma. Two stabilizing influences normally operate in a plasma with high electrical and thermal conductivity: (1) thermal conduction along magnetic field lines limits temperature gradients, and (2) the field is "frozen into" the plasma so reconnection cannot occur. The first stabilizing influence can be frustrated when the field is directed normal to the gradient of temperature. This allows a cool filament to form in the hot corona. The second stabilizer is disabled when the field is normal to the velocity gradient. This allows motion of the field relative to the plasma, and thus field reconnection. These theoretical situations have not yet been specifically identified in observational data.

As theorists recognized that strongly driven electric currents imply acceleration of charged particles, observers found various ways to detect these particles and to determine their numbers, energy, and motion. The next section describes some of these new observations and attempts to give some idea of the breakthroughs in understanding flares that can be expected when various aspects of the flare are viewed together and the major work of reconciling these views is accomplished.

Energetic Charged Particles

Energetic charged particles are now understood to carry a large proportion of the total energy in some (maybe all) flares. Energetic particles are detected and measured directly in space and indirectly by the radiation they emit or excite at the Sun and in the solar wind. Electrons, protons, heavier ions, and neutrons are detected directly by spacecraft-borne instruments that measure their energy, flux, speed, and direction, and determine the elemental and isotopic composition of the combined flux. This raises the expectation that conditions at the solar source could be deduced. In reality, some particle characteristics are so changed during the journey through the solar wind that their origin and initial characteristics are intricately intertwined with their propagation through a moving, structured magnetic plasma. Association with type II radio bursts suggests that solar energetic protons are accelerated in coronal shocks. A few cases give evidence that energetic charged particles observed in space are accelerated mainly in interplanetary shocks (Forman et al. 1982).

The remainder of this section describes radiation that the energetic particles emit in the radio spectral range and as X rays and gamma rays. Radio and X-ray data have long provided a steady flow of increasingly detailed information. Gamma radiation from flares, more recently observed, provides both new insight and useful constraints on solar-flare models.

Radio Bursts

Ground-based observations of radio emission at metric, then dekametric wavelengths gave some of the earliest data on the coronal aspect of flares. Well calibrated and complete observations at centimetric wavelengths provided a quantitative measure of flare size. Time profiles of flare bursts were classified, documenting the complexity and variety of flares. Space experiments extend the range of observation to low frequencies and the realm of observed sources out to the orbit of Earth. Direct observations confirm that the type III source is a "beam" of weakly relativistic electrons with energy of a few to about 100 keV. Fainberg and Stone's (1974) statistical analysis of the radio emission shows that the source electrons flow along the interplanetary magnetic field, which is itself pulled out by the solar wind and wound into a spiral by the Sun's rotation.

Observation of a type III burst (Fig. 2.12) is the culmination of a sequence of events. (1) Electrons are accelerated near an active region with strong magnetic fields. Where the magnetic structure is radial and "open," fast electrons escape in a directed beam. (2) The fastest electrons outrun the rest, and form a distinct group--a hump on the high-velocity tail of the velocity distribution. (3) In the ambient plasma, the electron stream excites plasma oscillations, or slowly propagating Langmuir waves at the plasma frequency, determined by the ambient electron density. (4) Plasma waves couple to electromagnetic waves, which are in the radio range, with frequency at the plasma frequency or its first harmonic. One condition that enhances coupling is a steep density gradient. (5) Radio waves can propagate to great distances if ambient density is low so that plasma frequency is lower than the radio-wave frequency. (6) Frequencies lower than about 10 MHz are

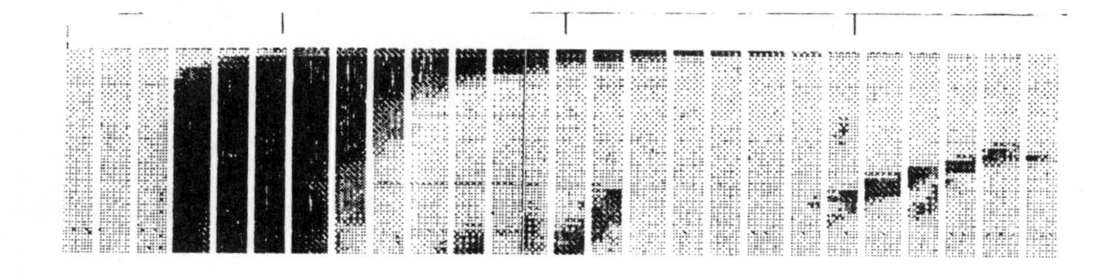

Figure 2.12--Type III and type II bursts observed by the Planetary Radio Astronomy Experiment on the Voyager spacecraft. The frequency scale is linear, from 1230 kHz at the bottom of the figure to 20 kHz at the top. Time markers at the top of the figure are separated by one hour.

reflected by the ionosphere and thus are unobservable from the ground. Space observatories measure solar radio bursts with frequencies as low as 30 kHz.

Parts (2) and (3) of this sequence were confirmed by a rare observation in space of plasma oscillations and of the velocity distribution of the responsible electrons (Lin et al. 1981). Analysis revealed the suspected "bump-on-tail" velocity distribution in convincing association with the simultaneous observation of plasma oscillations. Much remains to be learned about wave coupling, and particularly about propagation. The electrons that escape the Sun are probably a small fraction of those that are accelerated (Lin and Hudson 1976), yet the number of type III bursts far exceeds the number of flares, suggesting that charged-particle acceleration must be a common occurrence in active regions.

More rare is the observation of an interplanetary type II burst at frequencies below 1 MHz. Type II bursts at metric wavelengths have long been considered as evidence of shock waves propagating in the corona (Roberts 1959). Recent low-frequency observations from space follow the type II source for hours or days, and up to 1 a.u. from the Sun. When the source passes the spacecraft, simultaneous passage of a shock is recorded. Interplanetary type II bursts were detected at a rate of about one per month at maximum activity. They are not always clearly associated with a metric type II burst, but are preceded by an intense group of type III bursts. Cane et al. (1981, 1982) identified the type III precursors with metric fast-drift "herringbone structure," taken as evidence for shock acceleration of electrons.

Gamma Rays

The theory that describes gamma-ray emission by electrons, ions, and neutrons (Ramaty et al. 1975) was mostly in place by the time extensive observations of solar gamma-ray bursts became available. With theory already present, analysis of observations yielded an immediate harvest of information--some of it clarifying and some revolutionary.

An early result from the gamma-ray experiment (GRE) challenged the idea that "second-stage acceleration" is necessary to produce relativistic protons. One flare (1980 June 7) shows a series of seven bursts, with successive bursts separated by about 10 seconds. The same distinctive sequence is observed almost simultaneously in emission below

100 keV due to electrons, and in MeV photons produced by ions. The Mev ion emission lags the keV electron emission by less than a second. In this flare, there is no evidence of a separate, delayed mechanism to accelerate ions. Chupp (1982) summarized: "the rapidity with which ions and electrons accelerate . . . forces us to consider that a simple primary process is operative to energize both species."

Neutrons with energies of several hundred MeV were observed directly at the SMM spacecraft at times of two flares: 1980 June 21 and 1982 June 3. Energy of the neutrons was deduced from their arrival time at the spacecraft on the presumption that they were produced in a pulse at the Sun. In these two cases ions with energies ranging over more than 1 GeV appeared within a time interval of seconds, early in the impulsive phase. McDonald and Van Hollebeke (1985) pointed out that proton precursor events, associated with earlier flares, preceded the flares of 1980 June 7 and 21, and 1982 June 3. Thus pre-accelerated protons must have been present at the Sun when the protons and electrons of the main event were accelerated.

First GRE and then Hinotori found that gamma rays are emitted by many flares, some of them quite small. Summed spectra of weak gamma-ray line flares show enhanced features, not evident in the individual spectra, at frequencies of expected lines. This suggests that ion acceleration is often, if not always, present in flares. Surprisingly, white-light continuum emission does not appear to be a clear signature of ion acceleration (Chupp 1983).

Ramaty et al. (1983) found that observations strongly suggest that a single mechanism, operating under similar conditions, accelerates the protons and nuclei in all gamma-ray producing flares. This conclusion is based on comparison of disk and limb flares that leads to the inference that the shape of the energy spectrum of accelerated ions varies little from one flare to another.

However, Yoshimori et al. (1983) distinguished "impulsive" and "gradual" flares, presenting an example of each. In the gradual flare, gamma-ray line emission reaches maximum 40 seconds later than the maximum of hard X rays, whereas in the impulsive flare there is no such delay.

The energy-dependent peak delay of harder hard X rays behind softer hard X rays is one feature of gamma-ray line flares, according to Bai et al. (1983). They listed the following characteristics of flares with strong gamma-ray lines: (1) they are accompanied by intense radio bursts of types II and IV; (2) they are not necessarily large--some are subflares in H alpha; (3) they have flat hard X-ray spectra; and (4) high-energy (>300 keV) hard X-rays lag several seconds behind low-energy hard X-rays. Delay time seems to be correlated with flare size. Bai et al. noted that the presence of type II and IV bursts suggests acceleration at a shock. They proposed that protons, ions, and electrons all are energized by first-order Fermi acceleration operating in a flare loop.

Gamma-ray line emission, the signature of energetic ions in the photosphere, is not closely correlated with the flux of interplanetary energetic protons. Bai et al. (1985) studied a sample of gamma-ray/proton (GR/P) flares that produced gamma-ray lines, interplanetary protons, or both. As a group, gamma-ray line flares are associated with type II/type IV radio bursts, and with intense hard X rays with flat spectra that become flatter with time (spectral hardening or energy-dependent peak delay).

Among the flares with observed gamma-ray lines, Bai et al. found that those with long-duration (>90 minutes) hard X-ray spikes were clearly distinguished by a second characteristic, a high ratio of microwave (9 GHz) peak flux to hard X-ray peak flux. Then they studied a larger sample of flares with high microwave ratios as representative of gradual GR/P flares. These flares showed (1) long duration of soft X-ray and microwave emission, as well as hard X-ray emission; (2) large extent, in both H alpha and hard X rays; and (3) suggested association with interplanetary type II bursts, with coronal mass ejections, and with high fluxes of interplanetary energetic protons.

According to earlier analyses of hard X rays reported by Nitta et al. (1983) and Ohki et al. (1983), rarely occurring "gradual" flares are distinguished from "impulsive" flares by these characteristics: (1) long duration, greater than the 17 minutes allowed for a single flare by the tape recorder; (2) an extended source, high in the corona; (3) a power-law spectrum, in contrast to the exponential spectrum typical of the early stage of impulsive bursts; and (4) an energy-dependent delay of peak flux in which 300-keV photons reach maximum seconds or tens of seconds later than photons of 100 keV or less, so that the spectrum grows harder (flatter) with time.

Gamma-ray line emission indicates that energetic protons and ions are beamed down to the photosphere. In all gamma-ray line flares, protons are envisaged as accelerated in a closed magnetic loop. Protons and ions may be accelerated simultaneously with electrons, or may require preheating, perhaps at times by an earlier flare. In any case, they must stream down to the chromosphere and photosphere at the loop foot-points to produce gamma-ray line emission almost simultaneously with the hard X-ray burst.

The properties of gradual flares suggest that the source of their protons is higher in the corona, perhaps associated with a separate acceleration involving coronal shocks. The lag, in rare cases, of gamma-ray line emission behind hard X-ray emission suggests either second-step acceleration of protons and ions to energies beyond those attained in the first-stage acceleration shared by the electrons, or longer lifetime of protons and ions against energy loss to the ambient plasma. In the gradual flare, electrons are envisaged as trapped near the top of the loop, transferring their energy to the plasma of the low corona. The duration of emission from microwaves, X rays, and gamma rays is long enough to require a continued supply of accelerated electrons, protons, and ions.

Hard and Soft X Rays and Microwaves

Tanaka et al. (1982) noted that the rise of soft X-ray flux in flares parallels the time integral of the hard X-ray (>10 keV) flux; they proposed that soft X rays come from a plasma heated by collisions with the same population of energetic electrons responsible for the hard X rays. The three bursts they described are well-defined spikes. From the hard X-ray energy spectrum they derived the total energy of the source electrons (Hudson et al. 1978), and from the soft X-ray flux, the temperature, emission measure, and thermal energy content of the thermalized plasma. The energy of the soft X-ray emitting plasma and the energy of the hard X-ray emitting electrons are of comparable magnitude and show similar time variation during the rise and maximum of the bursts. This analysis provides support for electron-beam heating. It also determines density and temperature of the thermal plasma and, through a stability argument, ambient density.

A long-standing difficulty in interpreting hard X-ray and microwave fluxes may be resolved by a particularly simple argument. Although simultaneity of the X-ray and microwave emissions, down to minute detail, suggests that both originate from the same population of electrons, theoretical calculations predicted higher microwave fluxes than observed. Gary (1985) resolved the problem by postulating a thick-target model for the hard X rays and a realistically moderate magnetic field of a few hundred gauss within which the microwave-emitting electrons gyrate, producing a source that is optically thick at frequencies near and below 10 GHz. Earlier explanations of the apparent discrepancy are more complex; for example, Petrosian (1982) modeled microwave flux from the top of a magnetic loop and hard X-ray flux from footpoints with a population of energetic electrons spiraling with nearly isotropic pitch-angles in a loop with a small magnetic-field gradient.

These are just a few examples of how spatial, temporal, and spectral resolution in various experiments have improved understanding of the role of energetic charged particles in the sudden release of energy in magnetically confined plasma.

Is the spectral quality of flare radiation variable, and related to the ensuing geophysical effects? Can flares with most of their energy at the X-ray end of the spectrum be distinguished from those that produce shocks? Kahler (1982) noted that big flares are likely to be big in every way, showing major increases in all parts of the spectrum: big X-ray bursts, big radio bursts, and big proton events. Careful analysis is needed to find significant qualitative distinctions. The distinction between compact and two-ribbon flares, one of vertical extent, volume, duration, and association with coronal mass ejection, appeared as a clear separation in the work of Pallavicini et al. (1977), based on observations at a time of moderately low activity. It parallels the distinction between flares with low and high plasma beta, parallel and perpendicular currents. Sheeley et al. (1983), while confirming the qualitative trend of these relations, found a continuum of variation rather than distinct classes, possibly because burst duration is difficult to determine from global flux measurements made during high activity (Figure 2.13).

Analyses of observations with high resolution in space and time help define the physical conditions that explain the spectral quality of different flares. These studies show how the quality of radiation can vary from one flare to another. They explain differences in terms of the environment of the accelerated particles that transfer energy from the magnetic field to the plasma. If the particles are unobstructed until they slam into the dense chromosphere or transition region, hard X rays are emitted from a localized source at a footpoint of the magnetic loop. When density is enhanced throughout the loop so that the accelerated particles move through a denser plasma and release their energy higher up, in the corona, a softer spectrum is emitted from a larger, higher volume.

Tanaka et al. (1983) discussed two extreme prototypes

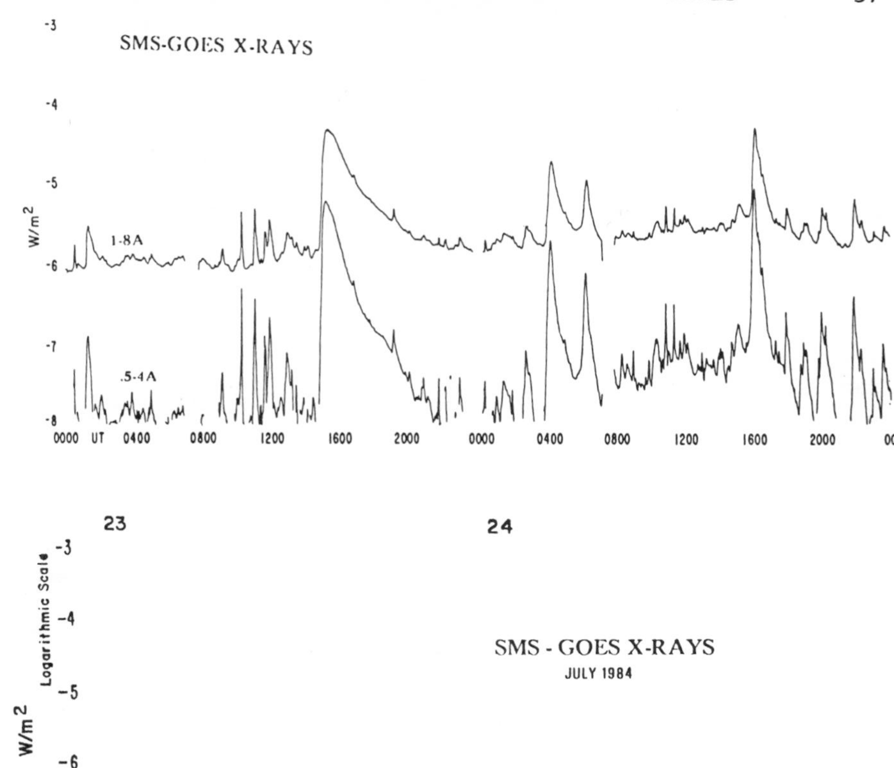

Figure 2.13--Time variation of soft X-ray flux at times of high and low activity.

distinguished by the relative importance of the impulsive hard X-ray burst and the smooth, long-duration soft X-ray burst. They suggested that the hard, spiky component is strong when the initiating electron beam impacts abruptly on the high-density transition layer after moving unimpeded through a thinly populated corona, and that the smooth, soft component predominates when the electrons give up most of their energy in a relatively dense corona over a relatively long path.

Tsuneta et al. (1984) called the latter, coronal type a

"hot thermal flare." They concluded that essentially all of the released energy goes immediately into heating the coronal plasma, and noted that these conditions are similar to the late phase of typical impulsive flares, but in the hot thermal flares density is high from the start. They inferred that preexisting high density leads to rapid transfer of energy to the ambient plasma so that unusually intense heating occurs from the beginning.

The high initial density that results in a hot thermal flare is linked to an earlier flare in a study of flare pairs by Strong et al. (1983). In each pair, the first X-ray flare is a fast hard spike located at a loop footpoint, whereas the second, probably in the loop top, has outstanding soft X-ray line emission. The second flare has the thermal spectrum of a particle population at fairly uniform energy rather than with energy concentrated in a few energetic electrons. In the first flare, a beam of energetic electrons is believed to impact on the chromosphere at the footpoints of a loop, producing microwave emission and hard X rays and heating the ambient material. This causes upward expansion of the heated plasma into the loop. This filling of the loop with dense plasma is often called "chromospheric evaporation." Attempts at more accurate descriptors include "ablation" and "convective flow," the filled loop becoming a "convective plume." When the loop is filled with dense material, particles newly accelerated near the loop top give up their energy in the loop rather than at the footpoints. The two peaks analyzed by Strong et al. (1983), separated by 2 to 4 minutes, would not be distinguished as separate events in most flare reports, so flare pairs could be more numerous than suspected.

Observations of source heights in a limb flare show how these ideas apply to different phases of a single impulsive flare (Simnett and Strong 1984). In the later part of the hard X-ray burst the source is observed to shift from the chromosphere into the corona, and the spectrum is seen to soften. Simnett and Strong concluded that "initially the flare energy is deposited below the transition zone, from which point the chromospheric plasma is ablated. When this plasma has expanded into the corona to the energy release site, particle acceleration associated with such energy release is quenched. The altitude of this site is estimated at 6000 $\pm$1500 km above the photosphere." They concluded also that their observations are consistent with the bulk of the energy being contained initially in suprathermal protons, in the energy range 0.1 to 1 MeV.

Kane (1983), in describing results of multi-spacecraft,

stereoscopic observations of high-energy (>100 keV) photon emission from five solar flares, concluded that (1) emission of the impulsive source probably originates in many separate non-thermal sources distributed in altitude, with low sources stronger; and (2) probably both trapped and precipitating electrons are present. Imaged X rays from Hinotori and from SMM show sources at heights as great as 40,000 km above the limb, often in close spatial and temporal relation to microwave sources.

A different process that leads to an elevated soft X-ray source involves accelerated charged particles and radio emission and propagation. In Melrose and Dulk's (1984) model of electron-cyclotron masers in flare loops, radio emission plays an unusually active role, transferring energy across magnetic lines of force. Weakly relativistic electrons (10 KeV to 1 MeV) trapped in a magnetic loop may mirror at the converging field lines at a foot of the arch or, if moving nearly parallel to the field lines, may penetrate to a region of high ambient density, collide, and give up their energy. The electron population thus assumes a "loss-cone distribution," depleted of electrons moving parallel to the field. Before leaving the loop, parallel-moving electrons share their energy with electromagnetic waves, which retain the energy because electrons with the amount and direction of motion necessary to take it back are missing from the population. The electromagnetic wave therefore grows rapidly. The resulting radiation, unlike the electrons, can escape the loop near the top and travel outward. Thus an appreciable amount of energy can be carried across the field of the flaring loop, and can cause localized heating and a large, high source of soft X-ray emission that is otherwise difficult to explain (Dulk 1985).

Melrose and Dulk (1984) suggested that the electron-cyclotron maser is capable of producing extremely fast, bright flare spikes in stars. Star flares are observed in X rays and at radio wavelengths. They add measurably to the total radiation from the star, so they are huge compared to solar flares. The study of star flares is sure to add new viewpoints to flare observations, and new challenges to flare theory.

A heirarchy of time and length scales were considered by Sturrock et al. (1985) who described distinct processes for different phases of a flare. For the gradual phase of energy release occurring at the beginning of most flares and at the end of some, they proposed a steady process of reconnection in a large magnetic structure in the corona, and for the impulsive phase, a more rapid stochastic process of reconnection. Conversion of magnetic energy from a single elementary flux tube was identified with the 2 to 10 second time scale of microwave radio elementary bursts. The flare impulsive phase, with a time scale of minutes, was attributed to sequential conversion of available energy from 100 to 1000 of these elements. For microwave sub-bursts, with a time scale of 0.1 second, they invoked "magnetic islands," substructures that form within elementary flux tubes in the course of the reconnection process.

2.5 FLARE PHYSICS AND FLARE PREDICTION

The pattern of large-scale velocity fields suggests the possibility of a long-term activity forecast. Howard and LaBonte (1983) said of their analysis of the torsional wave: "The most interesting result of this study is the possibility of predicting the onset of solar activity with a lead time of about a year. The amplitude increase of the westward flowing zones of the torsional wave can serve as a marker of the impending rapid rise of sunspot and magnetic activity on the surface. Close observations of the torsional wave during the next spot minimum will test whether this is a reliable harbinger of activity."

A more accurate mathematical description of the sunspot cycle should lead to more accurate prediction of future cycles if there really is mathematical regularity in the physical variation. Up to now, new data have usually failed to fit descriptions of old data.

Recognition that migration plays only a minor role in the large-scale distribution of magnetic flux is surely a breakthrough, but its immediate application to flare forecasting is not obvious. Perhaps only time (several decades) will tell whether these new physical insights lead to improvement over the present long-term forecasts based on the average character of the solar cycle.

Observation of flux cancellation offers an alternate process for the decay of the magnetic field in an active region. Moreover, studies of flux emergence and flux cancellation seem to offer new understanding of changing magnetic fields and of magnetic energy storage. So far, analysis of only one stage of one active region has been presented.

Energy storage in a stressed magnetic field is better understood and has been quantitatively estimated in a few cases. Observation of the complete magnetic vector improves this estimate. Is a continuously updated estimate of the flare potential of an active region almost within our grasp? How much improvement will it bring over the current estimates based on indirect clues?

High-resolution observations of flux emergence and of flux submergence or flux cancellation show suggestive association with chromospheric events in some cases; much more material must be analyzed before the promise of improved prediction may be realized.

The concept of magnetic energy storage and triggered release permeates the thinking of flare forecasters and of searchers for precursor triggers. Yet, at least two theorists emphasize that the main reservoir of stored energy may be hidden beneath the photosphere and the amount released may be only loosely related to the amount stored. The extent to which the release is stochastic is unknown.

Is the possibility of continuously driven magnetic reconnection a new insight or a red herring? This question bears directly on flare forecasting, but again, it is too soon to tell.

Hydrodynamic models of the impulsive phase show which processes are important and which can be ignored, simplified, or parameterized. They seem to provide a firm basis for increasingly complete translations of observations to physical conditions.

The model of the hot thermal flare and preconditioning by an earlier flare explains why the quality of X radiation can vary from flare to flare; if occurrence of successive flares in the same location could be foreseen, this would allow prediction of enhanced emission of soft X rays. In the near future, forecast users are likely to be asking for advice about the probable spectral quality of flare bursts, and flare forecasters may need to know the detailed history of each location. Forecasters may also learn to deduce the vertical structure and extent of active regions and to classify magnetic configurations in terms of parallel and perpendicular currents. Flare physics and flare forecasting are closely and intricately related through common questions. Constraints imposed by quantitative knowledge of radiation in different parts of the spectrum are sure to dispel misconceptions and improve flare models, but the precise quantitative knowledge necessary to improve practical forecasts still lies in the future.

Spicer and Brown (1981) remarked that "there exists no real model of a solar flare that can be used to compute the observed behavior of a flare, even given the initial conditions, boundary conditions, and pertinent parameters . . . (these conditions and parameters are currently unknown).

This situation has its origins in the complexity of the flare phenomenon, both observationally and theoretically, and the lack of the right kinds of data required to model such a situation." At present, flare theory is led by observation, and is more likely to benefit from the examined experience of forecasters than to provide useful tools for prediction. Nevertheless, theory does provide guidance in the search for significant parameters to observe and interpret. In his symposium summary, Brown (1983) made a plea that theoretical results should be given in a predictive fashion, rather than as an explanation through hindsight, and that theories should be simple, rather than try to explain all observational details.

2.6 REFERENCES

Akasofu, S.-I., 1985. Explosive magnetic reconnection: Puzzle to be solved as the energy supply process for magnetospheric substorms? EOS Trans. AGU **66**, 9.

Alfven, H., and P. Carlqvist, 1967. Currents in the solar atmosphere and a theory of solar flares. Solar Phys., **1**, 220-228.

Athay, R.G., 1981. The chromosphere and transition region, pp. 85-133 in The Sun as a Star, S. Jordan (ed.). Monograph series on nonthermal phenomena in stellar atmospheres, CNRS, NASA; NASA SP-450.

Babcock, H.W., 1961. The topology of the Sun's magnetic field and the 22-year cycle. Astrophys. J. **133**, 572-587.

Bai, T., B.R. Dennis, A.L. Kiplinger, L.E. Orwig, and K.J. Frost, 1983. Characteristics of gamma-ray line flares as observed in hard X-ray emissions and other phenomena. Solar Phys. **86**, 409-419.

Bai, T., A.L. Kiplinger, and B.R. Dennis, 1985. Two classes of gamma-ray/proton flares: impulsive and gradual. Rept. CSSA-ASTRO-85-15, Stanford Univ., Stanford, CA 94305.

Bracewell, R., 1985. Sunspot number series envelope and phase. Presented at Giovanelli Commemorative Colloquium, Part II, Tucson, Arizona, May; to be published in a special issue of Australian J. of Phys.

Brown, J., 1983. Conference summary, U.S.-Japan Seminar: Recent Advances in Understanding Solar Flares. Solar Phys. **86**, 458-460.

Bumba, V., and R. Howard, 1965. A study of the development of active regions on the Sun. Astrophys. J. **141**, 1492-1501.

Burlaga, L.F., and L.W. Klein, 1982. Interplanetary magnetic clouds at 1 a.u. J. Geophys. Res. **87**, 613-624.

Cane, H.V., R.G. Stone, J. Fainberg, R.T. Stewart, J.L. Steinberg, and S. Hoang, 1981. Radio evidence for shock acceleration of electrons in the solar corona. Geophys. Res. Lett. **8**, 1285-1288.

Cane, H.V., R.G. Stone, J. Fainberg, J.L. Steinberg, and S. Hoang, 1982. Type II solar radio events observed in the interplanetary medium I: General characteristics. Solar Phys. **78**, 187-198.

Canfield, R., C.-C. Cheng, K. Dere, G. Dulk, D. McLean, R. Robinson, Jr., E. Schmahl, and S. Schoolman, 1980. Radiative energy output of the 5 September 1973 flare, pp. 451-469 in Solar Flares, P. Sturrock (ed.). Colorado Assoc. Univ. Press, Boulder, Colo.

Chupp, E.L., 1982. Solar energetic photon transients (50 KeV -100 MeV). Proc. of La Jolla Inst. Workshop on Gamma-Ray Transients and Related Astrophysical Phenomena, U. of Calif. (San Diego), August 1981, AIP Conf. Proc. no. 77, Lingenfelter, R.E., H. Hudson, and D. Worrall (eds.), AIP, New York.

Chupp, E.L., 1983. High-energy particle acceleration in solar flares--observational evidence. Solar Phys. **86**, 383-393.

Cliver, E.W., L.C. Gentile, and J.M. Fink, 1982. Peak-flux-density spectra of large solar radio bursts. Bull. Am. Astron. Soc. **14**, 607.

Dicke, R., 1978. Is there a chronometer hidden deep in the Sun? Nature **276**, 676-680.

Dulk, G., 1985. Radio emission from the Sun and stars. Ann. Rev. of Astron. and Astrophys. **23**, 169-224.

Fainberg, J., and R.G. Stone, 1974. Satellite observations of type III solar radio bursts at low frequency. Space Sci. Rev. **16**, 145.

Forman, M.A., R. Ramaty and E.G. Zweibel, 1982. The acceleration and propagation of solar flare energetic particles. NASA TM 83989.

Gaizauskas, V., K.L. Harvey, J.W. Harvey, and C. Zwaan, 1983. Large-scale patterns formed by solar active regions during the ascending phase of cycle 21. Astrophys. J. **265**, 1056-1065.

Gary, D., 1985. The numbers of fast electrons in solar

flares as desuced from hard X-ray and microwave spectral data. BBSO #0240, submitted to Astrophys. J., in press.

Gurevich, A.V., and Ya.N. Istomin, 1979. Thermal runaway and convective heat transport by fast electrons in a plasma. Zh. Eksp. i Theor. Fiz. 50, 470-475; Soviet Physics--JETP **77**, 933-945.

Haurwitz, M., 1972. A current interruption, electrostatic discharge model of solar flares and magnetospheric substorms. Unpublished manuscript.

Hones, E.W., 1979. Transient phenomena in the magnetotail and their relation to substorms. Space Sci. Rev. **23**, 393-425.

Hones, E.W. (ed.), 1984. Magnetic reconnection in space and laboratory plasmas. Geophys. Monograph **30**, Am. Geophys. Union, Washington, D.C.

Howard, R., and B. LaBonte, 1983. The observed relationships between some solar rotation parameters and the activity cycle, pp. 101-111 in I.A.U. Symp. no. 102, Solar Magnetic Fields: Origins and Coronal Effects, J.O. Stenflo (ed.). Reidel, Dordrecht, Holland.

Hudson, H.S., R.C. Canfield, and S.R. Kane, 1978. Indirect estimation of energy deposition by non-thermal electrons in solar flares. Solar Phys. **60**, 137-142.

Hudson, H.S., and R.C. Willson, 1983. Upper limits on the total radiant energy of solar flares. Solar Phys. **86**, 123-130.

Kahler, S., 1982. The role of the big flare syndrome in correlations of solar energetic proton fluxes and associated microwave burst parameters. J. Geophys. Res. **87**, 3439-3448.

Kane, S.R., 1983. Spatial structure of high-energy photon sources in solar flares. Solar Phys. **86**, 355-365.

LaBonte, B., and R. Howard, 1982. Torsional waves on the Sun and the activity cycle. Solar Phys. **75**, 161-178.

Legrand, J.P., and P. Simon, 1981. Ten cycles of solar and geomagnetic activity. Solar Phys. **70**, 173-195.

Leighton, R.B., 1959. Observations of solar magnetic fields in plage regions. Astrophys. J. **130**, 366-380.

Leroy, J.-L., and J.-C. Noens, 1983. Does the solar-activity cycle extend over more than an 11-year period? Astron. Astrophys. **120**, L1-L2.

Lin, R.P., and H.S. Hudson, 1976. Non-thermal processes in large solar flares. Solar Phys. **50**, 153-178.

Lin, R.P., D.W. Potter, D.A. Gurnett, and F.L. Scarf, 1981. Energetic electrons and plasma waves associated with a solar type III radio burst. Astrophys. J. **251**, 364-373.

Livi, S., S. Martin, and J. Wang, 1985. The cancellation of magnetic flux on the quiet Sun. Presented at Giovanelli Commemorative Colloquium, part II, Tucson, Ariz.

Livingston, W., and J. Harvey, 1969. Observational evidence for quantization of photospheric magnetic flux. Solar Phys. **10**, 294-296.

McDonald, F., and M. Van Hollebeke, 1985. Helios I energetic particle observations of the solar gamma-ray/neuron flare events of 1982 June 3 and 1980 June 21. Astrophys. J. **290**, L67-L71.

McIntosh, P.S., 1981. The birth and evolution of sunspots: observations, pp. 7-57 in The Physics of Sunspots, L.E. Cram and J.H. Thomas (eds.). Sacramento Peak Natl. Obs. Conf., Sunspot, N.M., July 1981.

MacNeice, P., R.W.P. McWhirter, D. Spicer, and A. Burgess, 1984. A numerical model of a solar flare based on electron beam heating of the chromosphere. Solar Phys. **90**, 357-382.

Malitson, H., 1977. The solar spectrum, p. 26 in The Solar Output and Its Variation, O.R. White (ed.). Colorado Assoc. Univ. Press, Boulder, Colo.

Martin, S.F., and K.L. Harvey, 1979. Ephemeral active regions during solar minimum. Solar Phys. **64**, 93-108.

Martin, S., S. Livi, and J. Wang, 1985. The cancellation of magnetic flux II--in a decaying active region. Presented at Giovanelli Commemorative Colloquium, part II, Tucson, Ariz.

Melrose, D., and G. Dulk, 1984. Radiofrequency heating of the coronal plasma during flares. Astrophys. J. **282**, 308-315.

Nitta, N., T. Takakura, K. Ohki, and M. Yoshimori, 1983. Hard X-ray dynamic spectrum of flares observed by Hinotori. Solar Phys. **86**, 241-246.

Noyes, R.W., and E.J. Rhodes, Jr., 1984. Probing the depths of a star: the study of solar oscillations from space. NASA, JPL, Calif. Inst. of Technology, Pasadena, Calif.

Ohki, K., T. Takakura, S. Tsuneta, and N. Nitta, 1983. General aspects of hard X-ray flares observed by Hinotori: Gradual burst and impulsive burst. Solar Phys. **86**, 301-312.

Pallavicini, R., S. Serio, and G.S. Vaiana, 1977. A survey of soft X-ray limb flare images: The relation between their structure in the corona and other physical parameters. Astrophys. J. **216**, 108-122.

Petrosian, V., 1982. Structure of the impulsive phase of solar flares from microwave observations. Astrophys. J. **255**, L85-L89.

Priest, E., 1984. Magnetic reconnection at the Sun, in Magnetic Reconnection in Space and Laboratory Plasmas, E.W. Homes (ed.). Am. Geophys. Union, Washington, D.C., pp. 63-78.

Ramaty, R., R. Kozlovsky, and R.E. Lingenfelter, 1975. Solar gamma rays. Space Sci. Rev. **18**, 341-388.

Ramaty, R., R.J. Murphy, B. Kozlovsky, R.E. Lingenfelter, 1983. Gamma-ray lines and neutrons from solar flares. Solar Phys. **86**, 395-408.

Rieger, E., 1982. Gamma-ray measurements during solar flares with the gamma ray detector on SMM--an overview. Hinotori Symposium on Solar Flares, Inst. of Space and Astronautical Science, Tokyo, Japan, pp. 246-272.

Roberts, J., 1959. Solar radio bursts of spectral type II. Austral. J. Phys. **12**, 327-356.

Rosner, R., and G.S. Vaiana, 1978. Cosmic flare transients: Constraints upon models for energy storage and release derived from the event frequency distribution. Astrophys. J. **222**, 1104-1108.

Rust, D., 1976. Optical and magnetic measurements of the photosphere and low chromosphere. Trans. Roy. Soc. London A **281**, 427-433.

Rust, D., 1984. Permanent changes in filaments near solar flares. Solar Phys. **93**, 73-83.

Sawyer, C., 1967. A daily index of solar-flare activity. J. Geophys. Res. **72**, 385-391.

Sawyer, C., 1968. Statistics of solar active regions. Ann. Rev. Astron. and Astrophys. **6**, 115-133.

Sheeley, N.R., Jr., R.A. Howard, M.J. Koomen, D.J. Michels, 1983. Associations between coronal mass ejections and soft X-ray events. Astrophys. J. **272**, 349-354.

Shimabukuro, F.I., 1977. The solar spectrum above 1 mm, pp. 133-150 in The Solar Output and its Variation, O.R. White (ed.). Colorado Assoc. Univ. Press, Boulder, Colo.

Simnett, G.M., and K.T. Strong, 1984. The impulsive phase of a solar limb flare. Astrophys. J. **284**, 839-847.

Smith, D., and L. Auer, 1980. Thermal models for solar hard X-ray bursts. Astrophys. J. **238**, 1126-1133.

Smith, H., and E.v.P. Smith, 1963. Solar Flares. MacMillan Co., New York, and Collier-MacMillan Ltd., London.

Snodgrass, H., and R. Howard, 1985. Torsional oscillations of the Sun. Science **228**, 945-952.

Somov, B.V., S. Syrovatskii, and A. Spektor, 1981. Hydrodynamic response of the solar chromosphere to an elementary flare burst, I. Heating by accelerated electrons. Solar Phys. **73**, 145-155.

Somov, B.V., B. Sermulina, and A. Spektor, 1982. Hydrodynamic response of the solar chromosphere to an elementary flare burst, II. Thermal model. Solar Phys. **81**, 281-292.

Sonnerup, B., P. Baum, J. Birn, S. Cowley, T. Forbes, A. Hassam, S. Kahler, W. Matthaeus, W. Park, G. Paschmann, E. Priest, C. Russell, D. Spicer, R. Stenzel, 1984. Reconnection of magnetic fields, in Solar Terrestrial Physics: Present and Future, D.M. Butler and K. Papadopoulos (eds.). NASA Ref. Pub. 1120, pp. 1-3 to 1-42.

Spicer, D.S., and J.C. Brown, 1981. Solar flare theory, pp. 413-470 in The Sun as a Star, S. Jordan (ed.). Monograph series on nonthermal phenomena in stellar atmospheres, CNRS, NASA; NASA SP-450.

Steinberg, J.L., G.A. Dulk, S. Hoang, A. Lecacheux, and M.G. Aubier, 1984. Type III radio bursts in the interplanetary medium: The role of propagation. Astron. and Astrophys. **140**, 39-48.

Steinberg, J.L., S. Hoang, and G.A. Dulk, 1985. Evidence of scattering effects on the sizes of interplanetary type III radio bursts. Astron. Astrophys., in press.

Stewart, R., 1985. Solar noise storms and magnetic sector structures. Solar Phys. **96**, 381-395.

Strong, K.T., A.O. Benz, B.R. Dennis, J.W. Leibacher, R. Mewe, A.I. Poland, J. Schrijver, G. Simnett, J.B. Smith, Jr., and J. Sylwester, 1983. A multiwavelength study of a doubly impulsive flare. Solar Phys. **91**, 325-344.

Sturrock, P., 1966. Explosive and nonexplosive onsets of instability. Phys. Rev. Let. **16**, 270-273.

Sturrock, P.A., P. Kaufmann, R.L. Moore, and D.F. Smith, 1985. Energy release in solar flares. Solar Phys. **94**, 341-357.

Svestka, Z., 1976. Solar Flares. Dordrecht: D. Reidel.

Tanaka, K., N. Nitta, and T. Watanabe, 1982. The energetics of the elementary bursts. Hinotori Symposium on Solar Flares, Inst. of Space and Astronautival Sci., Tokyo, Japan, pp. 20-26.

Tanaka, K., N. Nitta, K. Akita, and T. Watanabe, 1983. Interpretation of the soft X-ray spectra from Hinotori. Solar Phys. **86**, 91-100.

Tokar, R.L., and D.A. Gurnett, 1980. The volume emissivity of type III radio bursts. J. Geophys. Res. **85**, 2353-2356.

Tsuneta, S., T. Takakura, N. Nitta, K. Ohki, K. Tanaka, K. Makishima, T. Murakami, M. Oda, Y. Ogawara, and I. Kondo, 1984. Hard X-ray imaging observations of solar hot thermal flares with the Hinotori spacecraft. Astrophys. J. **284**, 827-832.

Van Hoven, G., and G.J. Hurford, 1984. Flare precursors and onset. SMM workshop rept. (preprint).

Vernazza, J., E. Avrett, and R. Loeser, 1980. Structure of the solar chromosphere, III. Models of the EUV brightness components of the quiet Sun. Astrophys. J. Supp. **45**, 635-725.

Wang, J., H. Zirin, and Z.-X. Shi, 1984. The smallest observable elements of magnetic flux. BBSO #0239, Cal. Inst. of Technology, submitted to Solar Phys.

Warwick, J.W., 1962. The source of solar flares. Pub. Astron. Soc. of the Pacific **74**, 302-307.

Webb, D., C.-C. Cheng, G. Dulk, S. Edberg, S. Martin, S. McKenna-Lawlor, D. McLean, 1980. Mechanical energy output of the 5 September 1973 flare, pp. 472-499 in Solar Flares, P.Sturrock (ed.). Colorado Assoc. Univ. Press, Boulder, Colo.

Weber, R., 1978. Low frequency spectra of type III solar radio bursts. Solar Phys. **59**, 377-385.

Wild, J.P., S.F. Smerd, and A.A. Weiss, 1963. Solar bursts. Ann. Rev. Astron. Astrophys. **1**, 291-334.

Yoshimori, M., K. Okudaira, Y. Hirasima, and I. Kondo, 1983. Gamma-ray observations from Hinotori. Solar Phys. **86**, 375-382.

Zieba, St., M. Jaszczewska, and A. Michalec, 1982. The 810 MHz solar radio emission in the years 1968-1970. Acta Astron. **32**, 93-110.

Zirin, H., 1985. Evolution of solar magnetic fields. Presented at Giovanelli Commemorative Colloquium, part II, Tucson, Ariz.

3. *Flare Precursors*

3.1 INTRODUCTION

Some terrestrial and interplanetary effects of flares are delayed and thus can be usefully forecast even after the flare itself is well underway. Solar protons, for example, maximize an hour or many hours after the flare, and geomagnetic disturbance follows a day or so later. These effects can be predicted with more skill and confidence on the basis of an observed flare than on the basis of an expected flare. Heckman et al. (1984) stated: "Once a solar flare occurs, the capability to predict an ensuing proton event improves considerably."

For simultaneous flare effects--ionospheric effects of enhanced EUV and X radiation--similar forecasting skill would require knowledge of reliable precursors of flares. The need for accurate predictions of flares themselves and of their prompt effects drives the search for significant indicators that a flare is imminent. Study of flare precursors is further motivated by the expectation that knowledge of the environment, prehistory, and birth of flares will shed light on the whole flare process. This review of flare precursors emphasizes phenomena likely to aid predictability. Starting at the flare beginning and backing up in time, we will find scant evidence of verified predictability within time scales shorter than days or spatial scales much smaller than an active region. This review is thus complementary to that of Martin (1980a), which emphasized "distinct" precursors as an early stage in the physical development of a flare.

Schmahl (1983) compared the study of flare precursors to "exobiology, a subject in search of its subject matter." There is no shortage of omens, however, but rather a need for filters that can distinguish practical flare predictors from suggestive patterns and from activity that is only occasionally followed by a flare. To the reader who wants to become acquainted or reacquainted with chromospheric structures and their interpretation, we recommend a book: Nolan et al. (1970); two papers: Zirin and Tanaka (1973), Zirin (1984); and a report: Rust (1973). Each provides a logical presentation of well-illustrated examples of flare precursors. The SESC Glossary of Solar-Terrestrial Terms (1984) is particularly useful for nomenclature. Figure 3.1, reproduced from that source, shows many of the chromospheric features to be discussed.

3.2 FILAMENT ACTIVATION

Smith and Ramsey (1964), Martin and Ramsey (1972), and Martres et al. (1977) documented preflare changes in dark filaments seen in filtergrams that show different wavelengths within the broad profile of the H-alpha line. Line-of-sight velocities of tens of kilometers per second shift the profile in wavelength and cause complimentary changes of intensity in the filtergrams. Both upward and downward motions occur but the observers suggested that the most revealing preflare changes are blue shifts indicating upward velocities, toward the observer.

Dunn and Martin (1980) identified significant parameters for predicting filament eruption. The primary criterion is blue shift corresponding to outward velocity of at least 60 km/s. Another is observation of velocity shift over at least half the extent of the filament. Also important is the spatial pattern of Doppler shifts: either up one side of the filament and down the other, or up in the middle and down at both ends. When these conditions are fulfilled, the filament activation is termed "predictive." Almost two-thirds of the predictive filaments erupted within 24 hours of the prediction, whereas only 45% of the other filaments erupted. The exact relation of these filament eruptions to flares is not documented, although an association is implied.

Martin and Lawrence (1981) developed a quantitative comprehensive description of conditions influencing eruption of filaments outside new active regions. They found an enhanced rate of eruption among filaments close to the new active center and close to the time of birth of the new center. Rapid growth of the new region was found to enhance the chance of eruption, whereas a strong magnetic field

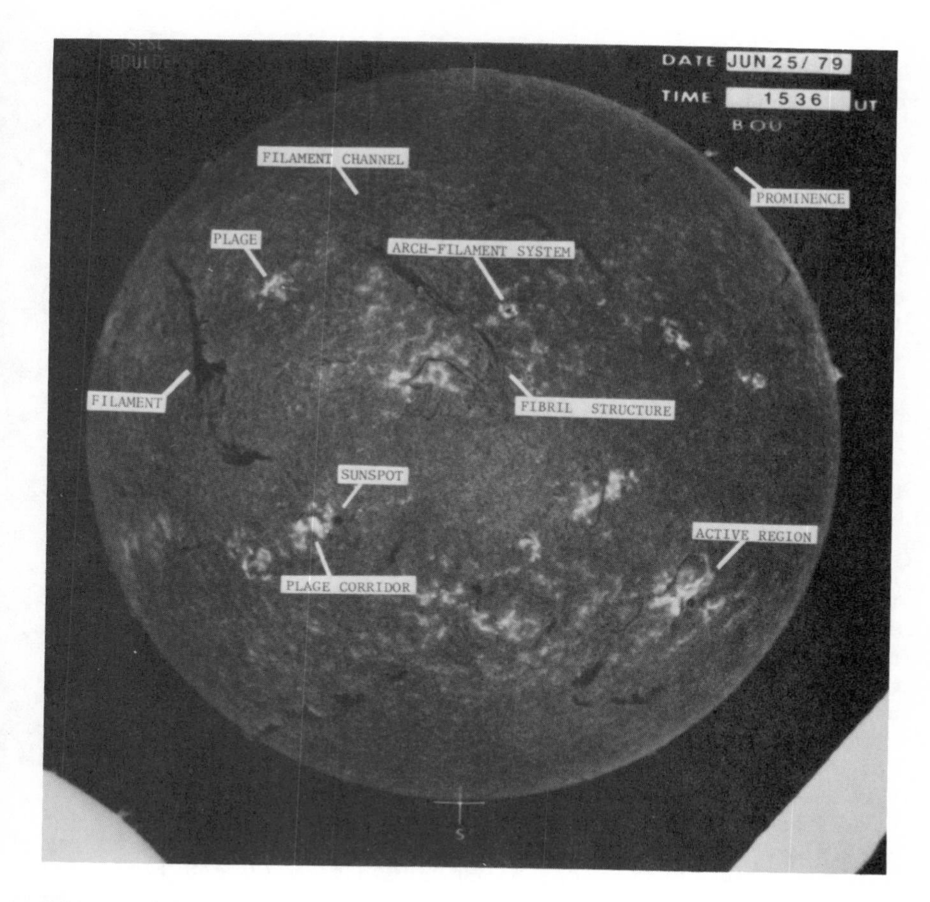

Figure 3.1--An H-alpha filtergram showing features of the solar chromosphere.

adjacent to the filament itself was found to be stabilizing.

These relations were applied in a practical forecasting experiment in July and August 1980 at solar-activity maximum. On the basis of active-region birth and growth, and magnetic-field strength estimated from plage brightness, the probability of eruption of quiescent filaments was forecast (Martin 1980b, Table IV). The tabulated results of this experiment permit a quantitative comparison between forecasts and actual eruption within 3 days of birth of the new active region. The comparison shows that the experiment can be considered unusually successful and the results particularly promising. The results are presented in terms of filament eruption; assessment of the value to flare forecasts will require clear definition of the number and size of flares that were anticipated and those that occurred.

A practical experiment in short-term forecasting of coronal mass ejections was carried out in 1980 during the Solar Maximum Mission. Observers at the San Fernando Observatory as well as at Mauna Loa and Sacramento Peak Observatories maintained an assiduous watch for filament and prominence eruptions and for filament activations. On several occasions response to a report of an eruption at or near the limb was rewarded by observation of a mass ejection in the corona. On the other hand, the rate of activations, generally Doppler shifts revealing large or changing line-of-sight velocities, was considerably higher than the rate of eruptions or of mass ejections so the proportion of false alarms was high. Because not all coronal mass ejections are flare related, the false alarm rate for flares must be even higher.

The relation of filament activity to flare occurrence was investigated by Neidig (1986) in a systematic study of 135 hours of simultaneous red-wing and blue-wing H-alpha observations with 1-minute time resolution. He recorded time and location of all sudden changes of absorption features as well as more gradual changes of "ominous size or appearance." This selection was judged to parallel that of a flare forecaster. The same filtergrams were then searched for flares, including flare-like brightenings smaller than those generally reported. The relatively stringent condition for association required the flare to be in contact with some part of the active filament. The observed interval from the start of filament activation to the beginning of the flare was 10 minutes or less in most cases, and Neidig concluded that associated preflare filament changes do not occur more than 30 minutes before the flare. The relative rates of filament activation and of flare brightening were found to depend on distance from the principal magnetic neutral line of the active region. Within 8000 km of this locus of polarity reversal, flares were found to outnumber filament activations but beyond that distance filament activations outnumbered flares by a factor of two. Near the neutral line, filament activation may be a reliable precursor--both of two cases were followed by flares--but it predicts only 10% of the flares. Farther from the neutral line, a large proportion, 79%, of the flares are preceded by filament activation, but 64% of all the filament activations give false alarms: they are followed by no flare-like brightening. This result for activations away from the neutral line is comparable to that of Martres et al. (1977).

Neither the size of the filament activation nor the importance of the flare-like brightening was found to affect the conclusions.

Significant differences of Neidig's study from studies that reach more positive conclusions may be (1) the requirement for close proximity of the flare to the filament or fibril, (2) consideration of all the material, negative cases as well as positive, and (3) the treatment of activations near and far from the neutral line as distinct situations.

An earlier study designed to test dependence of flare rate on filament activation and on X-ray brightening was carried out by Mosher and Acton (1980) who considered all filament activations and all flares within a limited period. They found no relation between filament activation and subsequent flares.

Thus studies of filament activation and flares present a mixed picture. Filament eruption is widely accepted as an indication of magnetic-field change and instability and as a primary and vital step in the flare process. Few argue against the concept that two-ribbon flares result from magnetic shear and reconnection and that filament eruption is the characteristic symptom of this rearrangement. The question is whether filament activity that precedes eruption and flares can be distinguished from filament activity followed by nothing more than minor brightening. Positive results will be convincing only when compared to false alarms, and negative results only when based on data that cover a spectral range wide enough to include Doppler-shifted moving material. Routine, automated observations are feasible (Martin 1980a), and filament activation remains near the top of the list of precursors deemed likely to be useful in forecasting flares.

3.3 PREFLARE BRIGHTENING

Three types of preflare brightening are distinguished here, although these distinctions may not be clear at all wavelengths:

1. Preheating: gradual, slight brightening at the flare site in the minutes preceding the impulsive flare or burst; the time profile of the X-ray burst may start with a low bump, interpreted as a slow increase in temperature and emission measure, that precedes the impulsive spike and eventually is buried by it.

2. Occurrence of bright points: in X-ray images, these are small enhancements adjacent to the site of the subsequent flare. H-alpha bright points may be very small flares or "microflares."
3. Plage brightening: region brightening and evolutionary brightening of the plage or part of the plage in various spectral regions.

Preheating within a coronal loop is described in Section 2.4 as a recent addition to understanding qualitative differences among flares. In practice it may be difficult to recognize those enhancements that precede flares and to distinguish them from those with no subsequent flare.

Lang (1980), Willson (1984), and Willson and Lang (1984) observed solar active regions at centimeter wavelengths with the Very Large Array, which resolves small regions of 15 arc sec or about 10,000 km. Circular polarization is interpreted as magnetic polarity. Brightness temperature of a million degrees identifies a coronal source. Observations covering more than 100 hours captured 3 flares. In the hour before each flare, and at no other time, changes of polarization were observed and interpreted as changes in magnetic-field configuration due to emerging flux (Lang 1980). In later observations at 2, 6, and 20 cm wavelengths reported by Willson (1983), only one of 8 bursts showed preheating. Although these results are disappointing in terms of flare prediction, the technique promises to be valuable in observing coronal magnetic fields.

A distinctive preflare signature in interferometric polarimetry at 10.6 GHz (3 cm wavelength) was described by Hurford and Zirin (1982, p. 17): An enhancement is "found to occur a few minutes to tens of minutes before the beginning of the impulsive phase of 11% of major solar flares. Physically, this signature is identified with the onset phase of a flare in which slow preheating of the active-region plasma may lead to the rapid release of magnetic energy in the impulsive phase. Although the signature can be quantified so as to permit its automated identification in real time, the false-alarm rate of 3 per week seriously compromises its suitability for practical flare prediction." Frequency-agile operation of the interferometer reveals that each preflare enhancement is confined to a narrow range of frequencies. This leads to the speculation that more preflare enhancements may be detected when a broad spectrum of frequencies is observed (Hurford et al. 1984). Sensitivity of microwave emission and propagation to density, temperature, and magnetic field will make interpretation of these observations a challenging task, but one that promises new information about coronal magnetic fields.

Harrison et al. (1985) identified small, discrete enhancements of imaged soft X-ray emission at the "departure time" of several flare-associated coronal mass ejections. These occur typically about 20 minutes before the flare itself. In some cases, the precursor bright feature moved slowly outward. The authors argued that even small enhancements are unlikely to occur by chance so close to the time and position of the flare. Sawyer (1986) showed that apparently early "departure time" can be interpreted as elevation of the source of the mass ejection. If precursor brightenings can be established as a significant early stage in the flare process, we will need to find out whether they can be distinguished from similar enhancements that are followed by no flare or mass ejection.

Van Hoven and Hurford (1985) noted that preheating should be observable in H alpha and in other wavelengths, and that "the quantitative study of such emissions, using modern photometric techniques, could exploit the high sensitivity and spatial resolution of modern filtergrams." A search of SOON H-alpha data is underway at NOAA for preflare pulsations or brightenings in a period of an hour or more before the flare (H. Kroehl, personal communication, 1984).

The second, flare-like, type of preflare brightening is often associated with filament activation, but is not identical. Intensified EUV "knots" were noted by Skylab astronauts as the features most likely to precede flares. The knots are bright also in X-ray lines and continuum and in H alpha (Machado and Noyes, 1978; Van Hoven et al. 1980). Cheng et al. (1982) observed UV bright points before flares, and noted that some are cospatial with UV flare kernels and some are not. They could not distinguish characteristic precursors from normal evolutionary change in the active region. Martin and Ramsey (1972), in their study of on-band and off-band H-alpha films made during the two hours preceding each of 297 large flares, found that the rate of brightenings doubled in the 15 minutes before the flare. These brightenings differed from subflares only in that they remained bright. Are these the same as the H-alpha bright points reported by the SOON network, which were found in numerical forecasting studies (Vecchia 1980, Neidig 1981) to supply predictive value, especially when they occurred along the neutral line?

Very faint, long-lived enhancements were discovered in X-ray images by Svestka and Schadee (1983). Some occurred at the location of a major flare during a period of about 32 hours preceding the flare, but a similar long-lived enhancement followed the flare; some of these HXIS-detected long-lived enhancements may be related to filament activation. Coronal structure in white light and X rays in the Skylab era are shown in Figure 3.2.

Figure 3.2--An image of coronal X-ray emission on the disk of the Sun superposed on an image of the white-light corona taken at eclipse. (The coronal image was made by a team from High Altitude Observatory, National Center for Atmospheric Research; the X-ray image and the composite are from American Science and Engineering, courtesy of Alan Krieger.)

The possibility that the current sheet between opposite magnetic polarities is marked by a structure of high density and low temperature that could be detected by microwave observations was discussed by Kuznetsov and Syrovatskii (1981).

There is negative evidence, as well, concerning a possible relation between X-ray brightening and subsequent flares: Kahler and Buratti (1976) failed to find convincing evidence of enhanced rate of X-ray brightening at the flare site in the 20 minutes before flares. Wolfson (1982) found X-ray preflare enhancements did not occur at the flare site, and Mosher and Acton (1980) found no evidence for a systematic or typical preflare change in X-ray brightenings, except when plage brightened.

Webb (telephone interview, 1984) suggested that negative results in the latter two cases may be explained by low spatial resolution, high background in the OSO-8 data used for these studies, and relatively low sensitivity to enhanced temperature. He extended the study of Kahler (1979) to a longer, 30-minute, preflare interval (Webb 1983) and found statistically significant enhancement of the occurrence rate of X-ray brightenings--bright X-ray loops, kernels, and sinuous features at locations adjacent to the flare site. Many of the preflare features occurred along the main neutral line, away from the flare site. Webb found no characteristics of the neighboring preflare X-ray enhancement that were correlated with the size of the subsequent "flares" (actually X-ray enhancements themselves, some very small).

Webb (1983) concluded that there was no preferred time for brightening to occur within the 30-minute interval examined. (We note, however, that if consideration is given to the number of images studied in each time interval, the occurrence histogram suggests that if plenty of images were available the rate would maximize 10 minutes before the flare.) In nearly all the cases of X-ray brightening for which H-alpha filtergrams were available, Webb observed simultaneous H-alpha brightening. He concluded that ". . . H-alpha emission is characteristic of preflare emission. However, searches for such preflare emission using H-alpha data alone will be difficult because the changes are subtle and don't always occur at the flare site." Thus Webb found H-alpha prebrightening less useful than X-ray precursors.

The yellow coronal emission line served in the 1950s as a useful index of activity (Dolder 1952). Altrock (1984) has resumed routine yellow-line observations with more sensitive equipment that detects the high-temperature emission at greater coronal heights. These data should be especially useful as an early indication of a very active region at east limb.

Table 3.1 provides a summary of some of the UV, X-ray, and radio-frequency flare precursors.

Flare frequency in 60 active regions was related to the region's brightness temperature at 32 GHz frequency (about 1 cm wavelength) by Kumagai and Ouchi (1986). They showed that the occurrence rate of X-ray bursts increases by a factor of about 10 when brightness temperature at 32 GHz increases by a factor of about 7.

Neidig (telephone interview, 1984) noted that this result may be expected simply from persistence of flare occurrence; radio brightness temperature of an active region may measure the flare rate in that region. Earlier studies of active-region microwave heating were reviewed by Bleiweiss et al. (1976) in their report of an effort to use microwave flux to forecast sudden ionospheric disturbance due to flares. They concluded that the method is "promising but far from providing definitive results."

Jackson (1979) found enhancement of the occurrence rate of metric type III fast-drift radio bursts several hours before flares. The type III rate has been used in practice in predicting ionospheric propagation conditions (Urbarz 1986). These are statistical relations, possibly the coronal counterpart of increasing activity in the region. Some type III bursts are individually flare related. Kane (1981) found that type III bursts occurring in time coincidence with X-ray bursts have special properties: they are intense and they start at high frequency--low in the corona, or in dense regions. Martres et al. (1977) searched for the optical counterpart of type III burst flare precursors and found small absorption features in off-band H-alpha images. Type III bursts that precede the type II, slow-drift bursts identified with flare-related shocks may also be distinctive: intense, occurring in groups, extending to low frequencies, and often accompanied by type V short-lived continuum. At the lowest frequencies, emitted in the high corona and in the solar wind, type III bursts are more readily identified with the herringbone structure of the accompanying metric type II than with metric type III bursts (Cane et al. 1981). Their excitation by streams of energetic electrons in the corona and their occurrence with the impulsive phase of the flare give type III bursts an important role in the search for understanding the whole process of flares and acceleration of energetic particles, but their

TABLE 3.1

SOME EUV, X-RAY, AND RADIO-FREQUENCY FLARE PRECURSORS

Precursor	Length Scale (arc sec)	Time Scale (min)	Location	Followed by Flare	Precedes Flare	Reference
EUV knots	5	5	near flare		8/12	van Hoven et al. 1980
X-ray >8A	>5	5-30	adjacent	16/22	16/23	Webb 1983
6-20 GHz	5-30	10-25			8/27	Hurford and Zirin 1982
					3/3	Lang 1980
					1/8	Willson 1983
					25%	Kai et al. 1983
H-alpha	10	5-10	adjacent			Martin 1980a
filament activation	300	0-30	adjacent	0-0.65	0-0.5	Martin 1980a, Neidig 1986

high occurrence rate reduces their value as practical precursors of individual flares.

To summarize, flare prebrightening has been investigated in many parts of the spectrum from X ray and EUV through H alpha to microwave radio. Although prebrightening has usually been studied independently at each wavelength, X-ray and H-alpha enhancements were determined to be cospatial and cotemporal with EUV enhancements. According to Webb (1984), because flares begin in the low corona, it is in the low corona that we should look for precursors. The basic input to flare predictions today is, in practice, coronal, in the form of "persistence" of occurrence of X-ray bursts. Flare-like bright points, especially those that occur along a neutral line, appear in numerical studies as a useful predictor. Is this another aspect of persistence? Actual preheating at the flare site that leads inevitably, or even probably, to development of a full-fledged flare has been pursued but, so far, has proved elusive.

3.4 MAGNETIC CONFIGURATION

Next to persistence, the index used most widely in short-term flare forecasting is the magnetic complexity of the active region as classified at Mount Wilson Observatory. Frequency of flare occurrence increases from alpha (unipolar) spots, through beta (simple bipolar group) to gamma (a complex group in which spots of opposite polarity are not neatly separable) (Giovanelli 1939). The Potsdam observers distinguished the delta configuration, in which spots of opposite polarity share the same penumbra (Kunzel 1965).

Warwick (1966) recognized the utility of the delta configuration as a predictor of major flares and flare effects, and Tanaka (1980) confirmed the close relation of the delta configuration to the occurrence of large flares. Avignon et al. (1964) distinguished "Configuration A"--a two-ribbon flare in a delta spot group--as very likely to produce energetic electrons. Investigations by Knoska and Krivsky (1983) and Klimes and Krivsky (1983) represent many studies showing that the delta configuration selects active regions that produce large H-alpha flares, radio bursts, X-ray bursts, sudden ionospheric disturbance, and radio bursts of type II (slow drift) and type IV (continuum). Although Achong and Stahl (1984) emphasized the stochastic component of flare production, their data show a strong dependence of flare rate on magnetic configuration (Sawyer et al. 1986).

Tang (1984) studied the origin of delta spots and found

that five of six cases were formed by the union of originally non-paired spots, shoved together by the birth or growth of neighboring spots. In the sixth case, the bipole emerged as a pair. In an earlier study, McIntosh (1969) concluded that in one case magnetic complexity and steep gradient resulted from sunspot growth, merging, and relative motion.

Ikhsanov (1982) related flare occurrence frequency to magnetic complexity arising from the birth of new spots within an existing sunspot group. Large flares are concentrated in complex spot groups with strong magnetic gradients that arise through birth of new spots (type I) or through proper motions and shear motions of neighboring bipoles (type II). Tang (1982) found that the high flare productivity of sunspot groups with reversed polarity is explained by the magnetic complexity of these regions.

Neutral-line complexity, measured by the number of kinks along the neutral line within the plage, was evaluated as a flare precursor by Lemmon (1970). A study of 250 flares that occurred in a 21-day period in 1969 showed that this criterion separated days of high flare index from days of lower flare index. A real-time forecasting experiment early in 1970 produced results "consistent with those of the sample population." Although this work has not been verified explicitly, numerical forecasting methods, described in Chapter 4, select as a useful predictor a variable based on the number of kinks in the neutral line.

Although significant H-alpha flares have been known to occur in decaying active regions after the sunspots have disappeared, such an event is rare indeed. Numerical studies and forecasters agree in finding the most reliable flare predictors to be active-region characteristics: known flare activity plus the nature of the magnetic field. Apparently significant are magnetic strength, gradient, rate of growth or decay, and complexity. These have been measured directly or estimated from sunspot observations. Another significant characteristic appears to be magnetic shear. Deduced in the past from velocity shear, sunspot motions, or orientation of H-alpha fibrils, magnetic shear can be measured directly, although with some uncertainty, as described in the next section.

3.5 MAGNETIC SHEAR

Knowledge of the three-dimensional vector magnetic field is generally considered to hold promise for short-term forecasting. This is one reason that solar physicists pursue the difficult measurement of the transverse field. In Section 5.3 we shall describe earlier work at the Crimean Astrophysical Observatory, where observation of the transverse field was directed specifically at predicting the occurrence of proton flares. The vector magnetograph developed at NASA's Marshall Space Flight Center provided routine observations for SMM in 1980 and has since been upgraded in sensitivity and signal-to-noise ratio (Hagyard et al. 1982). Analyses of the data yield the familiar magnetograms: mapped contours of strength and polarity that display strength, gradients, and neutral-lines of the longitudinal field. In addition, the vector magnetograph maps the transverse field and reveals magnetic shear. The transverse component of an unstressed potential field crosses the neutral line of the longitudinal field at a right angle (Fig. 3.3a). Thus, rotation of the transverse-field direction toward the direction of the neutral line (Fig. 3.3b) is a quantitative measure of shear and of stored energy in the magnetic structure of the active region.

Much of the interpretation of recent vector-magnetic data rests on the analysis of several days' observations of a single active region: AR 2372, 1980 April 5-7, the object of attention focused by the Flare Buildup Study (Svestka 1982). Hagyard et al. (1984) found that the degree of shear measured by the inclination of the transverse field to the neutral line maximized at two points on the neutral line and that flares began at these two points, A and B in Fig. 3.3d. At each of these sites there was a delta configuration as well as strong transverse field (Fig. 3.3c). Krall et al. (1982) found that pronounced velocity shear and tendency for alignment of the transverse field with the neutral line accompanied frequent and major flares and that relaxation of these indications of magnetic shear accompanied a decrease in the rate of flaring. Patty and Hagyard (1985) examined locations on the neutral line with delta configuration and with steep gradient of the longitudinal field. They found flare activity at these points to be related to the angle of shear measured by the inclination of the transverse field to the neutral line. Both strong gradient and strong shear appear to be necessary conditions for flare activity.

In the chromosphere-corona transition region, on the other hand, strongly sheared sites along the neutral line, identified by Athay et al. (1982), had little flare activity and corresponded to greatly reduced magnetic shear at the photospheric level. In the activity complex that they studied, flare occurrence was related rather to emerging magnetic flux. Hagyard (1984) supported the idea that it is the photospheric field and not the transition-region field that counts, noting that short H-alpha fibrils were aligned with the neutral line in flare-active situations whereas longer, presumably higher fibrils failed to show this alignment. A recent example of a change of fibril orientation at the time of a medium-sized flare was presented by Zirin (1984).

Vector magnetic fields have been measured in prominences (Athay et al. 1983, and references therein). The prominence field was found to be horizontal and inclined at almost any angle to the filament axis. Field strength, of the order of 15 gauss, varied little with height.

Buildup of magnetic, kinetic, thermal, and potential energy was calculated in a model of a sheared arcade of magnetic arches (Wu et al. 1984). Each arch crosses the neutral line and undergoes velocity shear that separates its feet in the direction along the neutral line (see Figure 3.4). Realistic initial values of linear scale, magnetic-field strength, and magnitude of velocity shear are based on values observed in AR 2372. The modeled change in magnetic energy exceeds the change in other forms of energy by a factor of 10. After a brief initial period of significant interaction among the various forms of energy, magnetic energy increased linearly with time in proportion to the shear or rate of displacement of the feet of each magnetic arch in the photosphere. In an arcade of moderate extent, energy storage can exceed the total energy released in flares, as estimated by Krall et al. (1982).

In a model of the evolution of a force-free magnetic loop, Su (1982) discovered, in addition to the solution describing expansion of the loop, a second solution describing compression. The compressed, sheared loop has strong electric currents and enhanced density and temperature. Su remarked that increasing the magnetic gradient and shear make the model similar to twisted prominences observed prior to large flares.

Results from the vector magnetograph are viewed hopefully because they give quantitative form to existing knowledge, both theoretical and practical. Only the magnetic field can store the amount of energy released in a flare. Comparison of the vector magnetic field to a potential field exhibits the stress in the observed field and indicates the amount of available energy. The observed and analyzed vector magnetic field has been compared to tested flare precursors--the magnetic delta configuration, velocity shear, and

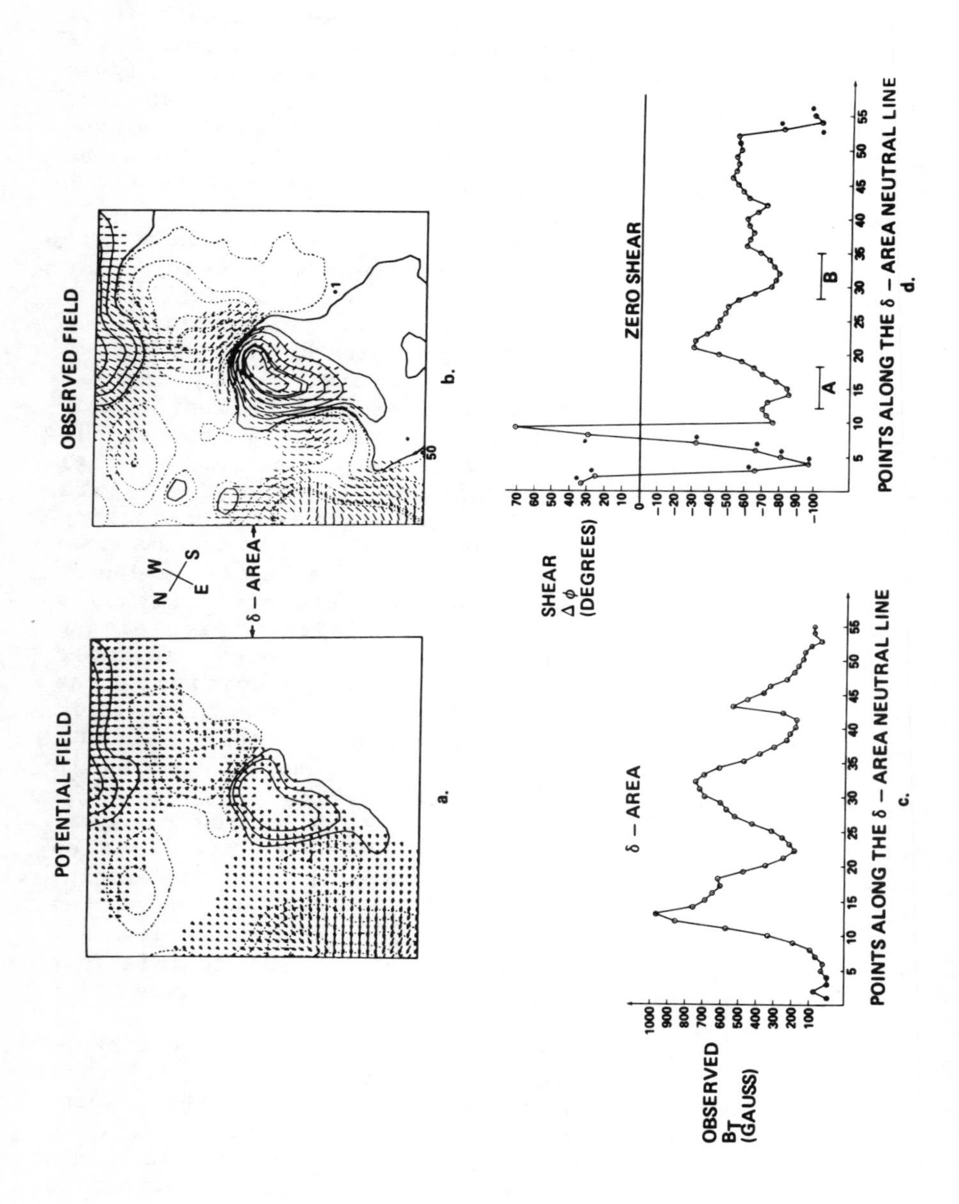
POTENTIAL FIELD
a.
OBSERVED FIELD
b.
N
W
S
E
δ – AREA
OBSERVED
B_T
(GAUSS)
δ – AREA
POINTS ALONG THE δ – AREA NEUTRAL LINE
c.
SHEAR
Δφ
(DEGREES)
ZERO SHEAR
A
B
POINTS ALONG THE δ – AREA NEUTRAL LINE
d.

Figure 3.3--Results from a vector magnetograph compared to calculated potential field. Direction and length of arrows represent direction and strength of the transverse magnetic field. Solid and dashed contours represent positive and negative polarities of the line-of-sight field. (a) Calculated potential field that fits the observed line-of-sight field. (b) Observed field. (c) Observed strength of the transverse component at each of 55 points along the neutral line (locations of points 1 and 50 are indicated in (b)). (d) Magnetic shear, defined as the angle between the observed transverse component and the calculated transverse component at each point. Points marked by asterisks represent values that are uncertain because the transverse field was weak. A and B mark points where H-alpha flare ribbons were most intense in a flare that occurred 7 hours earlier than these observations. These are also points where the transverse field was both strong and strongly sheared. Observations of AR2372, 1980 April 06, 2110 UT, (NASA, Marshall Space Flight Center). (These results are from Hagyard et al.(1984) and are reproduced with permission of D. Reidel Publishing Company.)

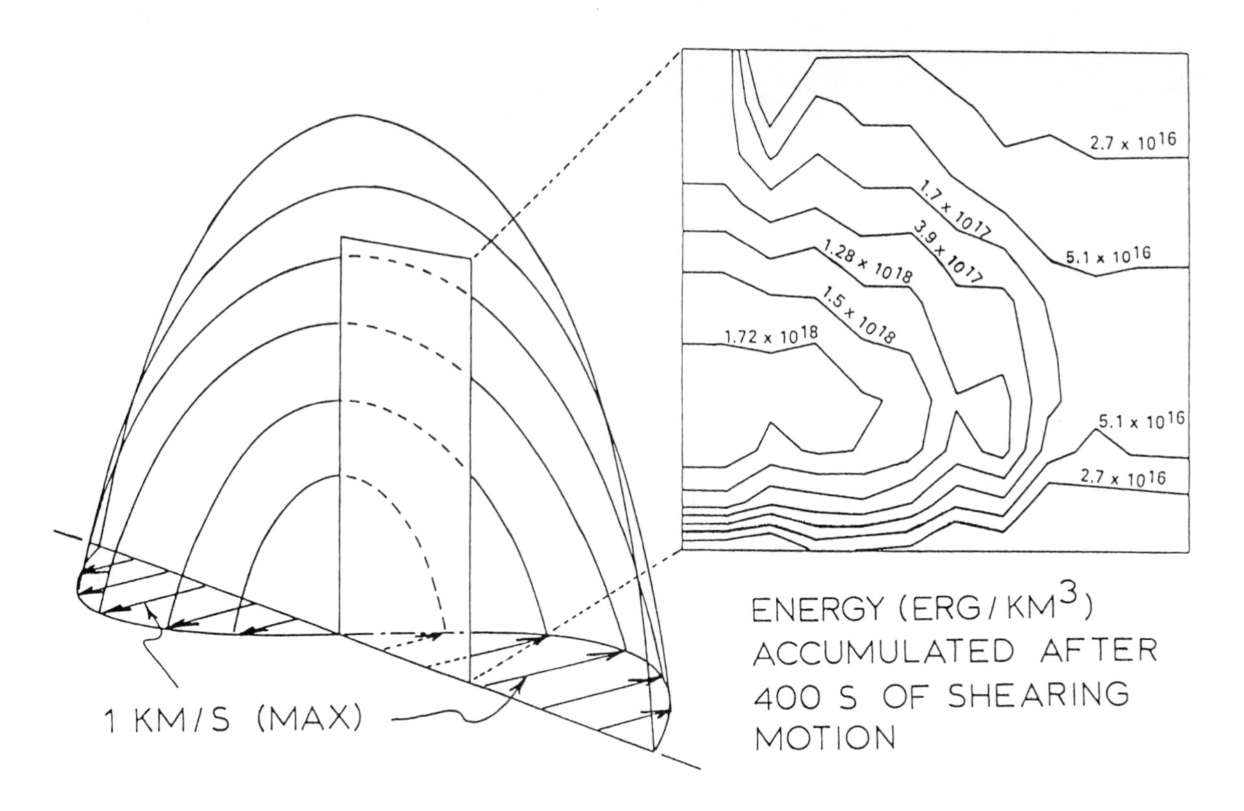

Figure 3.4—Magnetic shear and available energy in an active region. (Courtesy of Donald Neidig. Contours of accumulated energy were published in Wu et al. (1984) and are reproduced with permission of D. Reidel Publishing Company.

fibrils aligned along the neutral line--and it relates them quantitatively to magnetic shear.

3.6 VELOCITY FIELDS

Velocity and magnetic fields in the photosphere and chromosphere were studied by Harvey and Harvey (1976). Over a four-day period, 265 magnetograms of a flare-producing region were obtained. The observations include 14 subflares and flares. These were located near a magnetic neutral line with a considerable east-west component, and near a locus of reversal of the line-of-sight velocity field. For 8 of the 14 flares, a precursor velocity change was observed about a quarter of an hour before the flare start. This characteristic change was an increase amounting to about 0.5 km/s (always a blue shift, toward the observer) in a flare-sized area adjacent to the subsequent flare. The velocity change reached maximum near the time of flare intensity maximum, then decreased. Both this characteristic preflare change and the typical location of the flares with respect to the velocity field were observed only when the flaring region approached the limb. Harvey and Harvey concluded that the significant characteristic velocity field is parallel to the Sun's surface. The velocity "neutral line" is a line of divergence in the examples shown. This work, like the studies of vortical motions by Martres and colleagues (1973, 1974), suggests that horizontal velocity shear may be important to flare production. Harvey and Harvey (1976) noted also that temporal magnetic changes at the time of flares, although always consistent with the picture of relaxation of stress and decrease of field strength related to flare occurrence, hardly rise above the noise level of apparently random fluctuations. This agrees with Rust's (1976) result described in Section 2.3.

Athay et al. (1982) observed velocities in the transition region and found various orientations of the photospheric neutral line with respect to the direction of flow. Many studies have related sunspot motions to subsequent flaring in active regions. For flare forecasting, the significant features appear to be the result of these motions: shear and strong gradients in the magnetic field.

3.7 EMERGING AND CANCELLING MAGNETIC FLUX

Another recognized precursor of flare activity is emerging magnetic flux. Magnetograms show flux emerging into the photosphere as rapidly separating bipolar fragments. In the chromosphere, H-alpha filtergrams show brightening followed by the appearance of an arch-filament system (AFS): short, dark arches that connect opposite poles. Pores, then sunspots follow if the new emerging flux region (EFR) survives (Gaizauskas 1984). Important flares are noted when new flux emerges within an existing sunspot region, especially if this produces abnormal orientation of magnetic polarities. (But compare Tang's (1984) result, presented in Section 3.4, that complexity accounts for the apparent correlation of reversed polarity with enhanced flare rate.)

Sudden growth of magnetic complexity is consistent with the fast rates of flux appearance and of in situ disappearance found by Gaizauskas et al. (1983). A few active regions, especially those in one giant activity complex, account for a large proportion of the strong X-ray bursts that occurred in the period 1977 to 1979, according to Gaizauskas and McIntosh (1986). This agrees with the results of Bleiweiss et al. (1976) who noted that X-ray bursts in the period 1972 August through 1974 showed strong concentration in a few active regions, and commented that identification of these few very active regions would realize a good portion of the goal of forecasting major flares.

Neidig et al. (1978) studied 10 EFR in the context of flare activity and of other flare precursors. The 2 EFR that were definitely associated with flare activity arose in regions of magnetic shear, as deduced from H-alpha fibril orientation, and were further distinguished by large extent and large total magnetic flux. Neidig et al. estimated the stored magnetic energy, a function of total flux and shear angle and concluded that both emerging flux and magnetic shear are essential to flare activity. Sunspot motions and changes in sunspot area accompanied the flux increase and shear and could not be separated from them as possible flare precursors.

Martin et al. (1982) analyzed flares and emerging flux in a target region, Hale 16918, of the Flare Buildup Study. Of 17 EFR associated with new sunspots and with arch filaments, only 2 failed to produce flares, in contrast to the region studied by Neidig in which flares were attributed with certainty to only 2 of 8 EFR. Continued study of Hale

16918 (Martin et al. 1984) resulted in identification of EFR without flares, flares unrelated to EFR, and a variety of interactions. In the classic case, flares occur at the boundary between an EFR and old flux of opposite polarity. In other cases, emerging flux is observed to compress adjacent flux, steepening the gradient at a remote boundary, which then becomes a flare site. Alternatively, a flare may accompany eruption of a filament at such a remote boundary. The newly emerging flux is envisaged as connecting to the existing field, weakening the existing remote connecting magnetic structure that supported the filament.

Liggett and Zirin (1984) identified EFRs by appearance of AFS on high-resolution H-alpha filtergrams. They found the rate of flux emergence to be 27 times higher in active regions than in quiet regions. Furthermore, flux emergence is concentrated in some active regions more than in others. They identified flux emergence as the dominant means of creating the delta configuration.

Recent analysis at Big Bear Observatory of magnetograms and H-alpha filtergrams with high resolution in space and time emphasizes the significance of flux cancellation, defined by Martin et al. (1985) as "the mutual disappearance of magnetic flux in closely spaced features of opposite polarity." The term "cancellation" is preferred to "submergence" because it avoids any implication about the mechanism by which flux disappears. Flux cancellation involves opposite poles from different bipoles, for example, one pole of an EFR interacting with a fragment of opposite-polarity plage. Flux cancellation continues for several hours, during which time the magnetic gradient increases as the cancelling poles approach and collide. Flux disappears close to the polarity inversion line. Quantitative study of one decaying active region showed that cancellation accounted for essentially all flux loss. As the region decayed, emerging flux represented by each EFR resulted in net loss of flux, as the new poles interacted with existing fragments. Each of the 22 flares that occurred during the six-day study originated at a cancellation site. Some flares spread into regions where flux cancellation was not occurring and some cancellation events produced no flares. The time scale is hours for cancellation and minutes for flares. Because flux fragments of opposite polarity, once they begin to approach, invariably collide, the cancellation process is predictable on a time scale of hours (Martin et al. 1985).

A single case that involved emerging flux, preflare X-ray brightening near the flare site, and a postflare interregion X-ray arch was described by Sawyer et al. (1984) and Sawyer (1983). Slow evolution of large-scale coronal structure and continuum radio emission were observed well before the main flare and coronal mass ejection. The occurrence rate of this type of evolution when no flare follows was not estimated.

A case that lacked evidence of preflare emerging flux was described by Kundu et al. (1985). The observations were interpreted rather as showing rising and twisting of magnetic loops, and gradual evolution of large-scale magnetic patterns.

Because the subject arises persistently, we mention possible planetary influence on solar activity. In the 1960's, one of the authors investigated published and unpublished studies that related solar activity to some planetary attribute--for example, the motion of the Sun with respect to the center of mass of the solar system. The tidal effect of planets on the Sun was estimated to produce a change of about 1 mm in solar radius (1 part in 10^{12}). Positive conclusions in the literature were found to fall into three classes: (1) matching a period significant for activity with some combination of periods of planets and planetary satellites, without consideration of the enormous number of possible combinations; (2) comparison of a few 11-year cycles of activity to a parameter strongly influenced by Jupiter's 11.9-year period; (3) comparison of the rate of reported flares, which peaks at the northern-hemisphere summer solstice, to a planetary parameter that includes Earth. These analytical effects offered adequate explanation of most of the proposed planetary influences at that time.

3.8 SUMMARY

Solar active regions that are likely to flare are marked by magnetic complexity, including high gradients and shear. Characteristics such as these allow identification of flare-prone regions with time scales of the order of a day or more. The absence of such characteristics is the basis of a fairly reliable forecast of a flare-free period.

A practical test demonstrated the possibility of forecasting filament eruption on a 3-day time scale.

Prebrightening near the flare site within 30 minutes before the flare is just what is needed as a true precursor. Although there is evidence that flares are preceded by such prebrightening, no way has yet been found to distinguish the flare precursors from other small enhancements.

Van Hoven and Hurford (1985) noted that the energetic processes at work before a flare span the temperature range from 10 thousand to 100 million degrees, and can be observed completely only by using a variety of instruments. They stated, "Although each of the resulting data sets may be well analyzed, any individual instrument merely provides a look at part of the flare. Comprehensive, multi-wavelength, joint analyses of preflare conditions are difficult and rare." Perhaps we should not be dismayed that the first few attempts have not solved all problems.

Whether early recognition of an imminent flare or prediction of the time of the next flare will eventually be possible is not known, but the sources of such information that presently seem to be most likely are filament activation, magnetic-flux evolution, and preheating, monitored over a wide range of frequencies.

3.9 **REFERENCES**

Achong, A., and P. Stahl, 1984. SID flare production. Solar Phys. **92**, 259-269.

Altrock, R., 1986. Coronal emission-line data and solar-terrestrial predictions. Proc. of Meudon Solar-Terrestrial Predictions Workshop (1984), G. Heckman, M. Shea, and P. Simon (eds.), Boulder, Colo.

Athay, R.G., J. Gurman, W. Henze, and R. Shine, 1982. Fluid motions in the solar chromosphere-corona transition region. II. Active-region flows in CIV from narrow slit Dopplergrams. Astrophys. J. **261**, 684-699.

Athay, R.G., C. Querfeld, R. Smartt, E. Landi Degl'Innocenti, and V. Bommier, 1983. Vector magnetic fields in prominences. Solar Phys. **89**, 3-30.

Avignon, Y., Martres, M.-J., and Pick, M., 1964. Identification de classes d'eruptions chromospheriques associees aux emissions de rayons cosmiques et a l'activite radioelectrique. Ann. d'Ap. **27**, 23-28.

Bleiweiss, M., F. Wefer, and M. Hurst, 1976. Analysis of radiospectroheliograms at wavelengths of 8.6 mm and 2.0 cm in predicting communication outages. NELC, Tech. Note 3262.

Cane, H.V., R. Stone, J. Fainberg, R. Stewart, J.L. Steinberg, and S. Hoang, 1981. Radio evidence for shock acceleration of electrons in the solar corona. Geophys. Res. Let. **8**, 1285-1288.

Cheng, C.-C., E. Bruner, E. Tandberg-Hanssen, B. Woodgate, R. Shine, P. Kenny, W. Henze, and G. Poletto, 1982.

Observations of solar flare transition zone plasmas from the Solar Maximum Mission. Astrophys. J. **253**, 353-366.

Dolder, F., 1952. Solar activity and the yellow coronal line 5694 A. Ph.D. thesis, Dept. of Physics, Univ. of Colorado.

Dunn, J.M., and S. Martin, 1980. An attempt to identify flare precursor mass motions in real time. Bull. Am. Astron. Soc. **12**, 904.

Gaizauskas, V., 1982. The relation of solar flares to the evolution and proper motions of magnetic fields. Adv. in Space Res. **11**, 11-30.

Gaizauskas, V., 1984. Emerging flux. Submitted to Proc. of the NASA-SMM Workshop on Solar Flares.

Gaizauskas, V., K. Harvey, J. Harvey, and C. Zwaan, 1983. Large-scale patterns formed by solar active regions during the ascending phase of cycle 21. Astrophys. J. **265**, 1056-1065.

Gaizauskas, V., and P. McIntosh, 1986. On the flare effectiveness of recurrent patterns of magnetic fields. Proc. of Meudon Solar-Terrestrial Predictions Workshop (1984), G. Heckman, M. Shea, and P. Simon (eds.), Boulder, Colo.

Giovanelli, R., 1939. The relations between eruptions and sunspots. Astrophys. J. **89**, 555-567.

Hagyard, M.J., 1984. The relation of sheared magnetic fields to the occurrence of flares. Invited lecture presented at Solar-Terrestrial Prediction Workshop, Meudon, 1984 June 20, to be published in Artificial Satellites (Biuletyn Polskich Obserwacji Sztucznych Satelitow), a publication of the Polish Acad. Sci., Natl. Com. on Geophys.

Hagyard, M.J., N.P. Cumings, E.A. West, and J.B. Smith, Jr., 1982. The MSFC vector magnetograph. Solar Phys. **80**, 33-51.

Hagyard, M.J., J.B. Smith, Jr., D. Teuber, and E.A. West, 1984. A quantitative study relating observed shear in photospheric magnetic fields to repeated flaring. Solar Phys. **91**, 115-126.

Harrison, R., P. Waggett, R. Bentley, K. Phillips, M. Bruner, M. Dryer, and G. Simnett, 1985. The X-ray signature of solar coronal mass ejections. Solar Phys. (in press).

Harvey, K.L., and J.W. Harvey, 1976. A study of the magnetic and velocity fields in an active region. Solar Phys. **47**, 233-246.

Heckman, G., J. Hirman, J. Kunches, and C. Balch, 1985. The

monitoring and prediction of solar particle events--an experience report. Advances in Space Res. **4**, 165-172 (journal of COSPAR).

Hurford, G., and H. Zirin, 1982. Interferometric observations of solar flare precursors at 10.6 GHz. AFGL-TR-82-0117, Hanscom AFB, Mass.

Hurford, G., R. Read, and H. Zirin, 1984. A frequency-agile interferometer for solar microwave spectroscopy. BBO 0224, to be published in Solar Phys.

Ikhsanov, R.H., 1982. Configuratsii magnitnovo polya v gruppakh pyaten i solnechnie vspuishki. I, Izv. Glavnoi Astronomicheskoi Obs. b Pulkove, no. 200, Astrofiz. i Astrometria.

Jackson, B.V., 1979. Evidence for a peak in the number of isolated type III bursts prior to large solar flares. Proc. Astron. Soc. of Australia **3**, 383-386.

Kahler, S., 1979. Preflare characteristics of active regions observed in soft X-rays. Solar Phys. **62**, 347-357.

Kahler, S., and B. Buratti, 1976. Preflare X-ray morphology of active regions observed with the AS&E Telescope on Skylab. Solar Phys. **47**, 157-165.

Kai, K., H. Nakajima, and T. Kosugi, 1983. Radio observations of small activity prior to main energy release in solar flares. Pub. Astron. Soc. Japan **35**, 285-298.

Kane, S.R., 1981. Energetic electrons, type III radio bursts, and impulsive solar-flare X rays. Astrophys. J. **247**, 1113-1121.

Klimes, J., and L. Krivsky, 1983. Occurrences of flares with type II and IV radio events in interacting sunspot groups in the course of revolutions. Bull. Astron. Inst. Czechosl. **35**, 273-275.

Knoska, S., and L. Krivsky, 1983. Flares of type II and IV radio bursts in magnetic types of sunspot groups. Pub. Debrecen Heliophysical Obs. **5**, 557-566.

Krall, K.R., J.B. Smith, Jr., M.J. Hagyard, E.A. West, and N.P. Cumings, 1982. Vector magnetic field evolution, energy storage, and associated photospheric velocity shear within a flare-productive active region. Solar Phys. **79**, 59-75.

Kumagai, H., and C. Ouchi, 1986. 32 GHz solar radio observation at Hiraiso, Japan and its application to short term solar activity prediction. Proc. of Meudon Solar-Terrestrial Predictions Workshop (1984), G. Heckman, M. Shea, and P. Simon (eds.), Boulder, Colo.

Kundu, M.R., V. Gaizauskas, B. Woodgate, E. Schmahl, R.

Shine, and H.P. Jones, 1985. A study of flare buildup from simultaneous observations in microwave, H alpha, and UV wavelengths. Astrophys. J. Supplement **57**, 621-630.

Kunzel, H., 1965. Zur klassifikation von Sonnenflecken-gruppen. Astron. Nach. **288**, 177-181.

Kuznetsov, V., and S. Syrovatskii, 1981. On the possibility of observation of current sheets in radio band. Solar Phys. **69**, 361-372.

Lang, K.R., 1980. Solar flare prediction using radio wave-length interferometers, pp. C-131-143 in Solar Terrestrial Predictions Proceedings, vol. III, Solar activity predictions (R.F. Donnelly, ed.). U.S. Dept. Commerce, NOAA, ERL, Boulder, Colo. (Supt. Doc., U.S. Govt. Printing Off., Washington, D.C. 20402).

Lemmon, J., 1970. Forecasting flares from inferred magnetic fields. Presented as paper 70-1372 at the AIAA Observation and Prediction of Solar Activity Conf., Huntsville, Ala.

Liggett, M., and H. Zirin, 1984. Emerging flux in active regions. Solar Phys. **97**, 51-58.

Machado, M., and R. Noyes, 1978. Lyman continuum observations of solar flares. Solar Phys. **59**, 129-140.

McIntosh, P., 1969. Sunspots associated with the proton flare of 23 May 1967, pp. 14-19 in Data on solar event of May 23, 1967 and its geophysical effects, J.V. Lincoln (ed.). Rept. UAG-5, ERL, ESSA, Boulder, Colo.

Martin, S., 1980a. Preflare conditions, changes and events. Solar Phys. **68**, 217-236.

Martin, S., 1980b. Observations before and during the flash phase of solar flares. Ann. Tech. Rept. IV, AFOSR Contract F49620-78-C0025, Calif. State Univ. Foundation, Northridge, Calif.

Martin, S., and V. Lawrence, 1981. A formula for forecasting the probability of eruption of quiescent filaments. Bull. Am. Astron. Soc. **13**, 847.

Martin, S.F., and H.E. Ramsey, 1972. Early recognition of major solar flares in H alpha, pp. 371-387 in Solar Activity, Observations and Predictions, P. McIntosh and M. Dryer (eds.). MIT Press, Cambridge, Mass.

Martin, S., L. Deszo, A. Antalova, A. Kucera, and K.L. Harvey, 1982. Emerging magnetic flux, flares and filaments--FBS interval 16-23 June 1980. Adv. in Space Res. **2**, 39-52.

Martin, S., R. Bentley, A. Schadee, A. Antalova, A. Kucera, L. Deszo, L. Gesztelyi, K. Harvey, H. Jones, S. Livi, and J. Wang, 1984. Relationships of a growing magnetic flux region to flares. Adv. in Space Res. **4**, 61-70.

Martin, S., S.H.B. Livi, and J. Wang, 1985. The cancellation of magnetic flux II--in a decaying active region. To be published in the Australian J. of Physics, Ron Giovanelli commemorative volume.

Martres, M.-J., 1974, in Flare-related magnetic field dynamics. Conf. Proc. High Altitude Obs., p. 333.

Martres, M.-J., I. Soru-Escaut, and J. Rayrole, 1973. Relationship between some photospheric motions and the evolution of active centers. Solar Phys. **32**, 365-378.

Martres, M.-J., I. Soru-Escaut, and Y. Nakagawa, 1977. H-alpha offband preflare activities. Astron. & Astrophys. **59**, 255-259.

Mosher, J., and L. Acton, 1980. X-rays, filament activity and flare prediction. Solar Phys. **66**, 105-111.

Neidig, D.F., 1986. H-alpha filament and fibril activity as a short-term (30-minute) predictor of flares and flare-like events. Proc. of Meudon Solar-Terrestrial Predictions Workshop (1984), G. Heckman, M. Shea, and P. Simon (eds.), Boulder, Colo.

Neidig, D.F., H.L. DeMastus, and P.H. Wiborg, 1978. Flares, force-free fields, emerging flux, and other phenomena in McMath 14943 (September 1977). Rept. AFGL-TR-78-0194, Envir. Res. Papers, No. 637, AFGL, Hanscom AFB, Mass.

Neidig, D., P. Wiborg, P. Seagraves, J. Hirman, and W, Flowers, 1981. An objective method for forecasting solar flares. AFGL-TR-8-0026, Air Force Geophysics Lab., Hanscom AFB, Mass.

Nolan, B., S. Smith, and H. Ramsey, 1970. Solar filtergrams of the Lockheed Solar Observatory. Lockheed Solar Obs., Palo Alto, Calif.

Patty, S., and M. Hagyard, 1985. Delta configuration: Flare activity and magnetic-field structure. Solar Phys., in press.

Rust, D., 1973. Estimating the flare-production potential of solar active regions from analysis of real-time magnetic field data. AFCRL-TR-73-0221.

Rust, D., 1976. Optical and magnetic measurements of the photosphere and low chromosphere. Trans. Roy. Soc. London A **281**, 427-433.

Sawyer, C., 1983. The 1980 April 12 flare and transient: Report on progress in interpretation. Adv. in Space Res. **2**, 265-270.

Sawyer, C., 1986. Coronal mass ejection in the flare time sequence. Proc. of Meudon Solar-Terrestrial Predictions Workshop (1984), G. Heckman, M. Shea, and P. Simon (eds.), Boulder, Colo.

Sawyer, C., G. Simnett, F.T. Erskine, III, and T. Gergely, 1984. Solar and interplanetary observations of the 1980 April 12 west-limb flare, pp. 237-252 in STIP Symposium on Solar/Interplanetary Intervals, M.A. Shea, D.F. Smart, and S.M.P. McKenna-Lawlor (eds.). Engineering Internatl., Inc., Huntsville, Ala.

Sawyer, C., A. Achong, and P. Stahl, 1986. SID flare production and Mt. Wilson magnetic class: An alternative interpretation. Solar Phys., in press.

Schmahl, E.J., 1983. Flare buildup in X-rays, UV, microwaves and white light. Adv. in Space Res. **2**, 73-90.

SESC Glossary of Solar-Terrestrial Terms, 1984. Space Environment Services Center, NOAA/ERL R/E/SE2, Boulder, Colo.

Smith, S.F., and H.E. Ramsey, 1964. The flare-associated filament disappearance. Zs. fur Astrophys. **60**, 1-18.

Su Q.-R., 1982. Flare build-up in preflare magnetic loops and nonlinear force-free magnetic fields. Solar Phys. **75**, 229-236.

Svestka, Z., 1982. Flare build-up study in the SMA period. Adv. in Space Res. **11**, 3-4.

Svestka, Z., and A. Schadee, 1983. Pre- and post-flare X-ray variations in active regions. Solar Phys. **86**, 267-277.

Tanaka, K., 1980. Prediction of some great flkares based on the magnetic field configuration and evolution of sunspot groups, pp. C-1-11 in Solar Terrestrial Predictions Proceedings, vol. III, Solar activity predictions (R.F. Donnelly, ed.). U.S. Dept. Commerce, NOAA, ERL, Boulder, Colo. (Supt. Doc., U.S. Govt. Printing Off., Washington, D.C. 20402).

Tang, F., 1982. Reversed-polarity regions. Solar Phys. **75**, 179-183.

Tang, F., 1984. On the origin of delta spots. Solar Phys. **89**, 43-50.

Urbarz, H., 1986. Occurrence frequency of solar type III bursts. Proc. of Meudon Solar-Terrestrial Predictions Workshop (1984), G. Heckman, M. Shea, and P. Simon (eds.), Boulder, Colo.

Van Hoven, G., and G. Hurford, 1985. Flare precursors and onset. Report to FBS Workshop, Dept. of Physics, Univ. of California, Irvine, Calif.

Van Hoven, G., U. Anzer, D. Barbosa, J. Birn, C.-C. Cheng, R. Hansen, B. Jackson, S. Martin, P. McIntosh, Y. Nakagawa, E. Priest, E. Reeves, E. Reichmann, E. Schmahl, J. Smith, C. Solodyna, R. Thomas, Y. Uchida, and A. Walker, 1980. The preflare state, pp. 18-81 in Solar Flares, P. Sturrock (ed.). Colo. Assoc. Univ. Press, Boulder, Colo.

Vecchia, D., P. Tryon, G. Caldwell, and R. Jones, 1980. Statistical methods for solar flare probability forecasting. Rept. AFGL-TR-0-0336, Air Force Geophysics Lab., Hanscom AFB, Mass.

Warwick, C., 1966. Sunspot configurations and proton flares. Astrophys. J., **145**, 215-223.

Webb, D.F., 1983. A study of coronal precursors of solar flares. AFGL-TR-83-0126, Hanscom AFB, Mass., and ASE-4874 and submitted to Solar Phys.

Webb, D., 1984. A review of collective or statistical studies of coronal flare precursors. Rept. to FBS Workshop, Study Group D on Flare Precursors and Onset, COSPAR XXV.

Willson, R.F., 1983. High-resolution observations of solar radio bursts at 2, 6, and 20 cm wavelength. Solar Phys. **83**, 285-303.

Willson, R.F., 1984. Observations of preburst heating and magnetic field changes in a coronal loop at 20 cm wavelength. Solar Phys. **92**, 189-198.

Willson, R.R., and Lang, K.R., 1984. Very Large Array observations of solar active regions. Astrophys. J. **279**, 427-437.

Wolfson, C.J., 1982. Soft X-ray emission from active regions shortly before solar flares. Solar Phys. **76**, 377-386.

Wu, S.T., Y.Q. Hu, K.R. Krall, M.J. Hagyard, and J.B. Smith, Jr., 1984. Modeling of energy buildup for a flare-productive region. Solar Phys. **90**, 117-131.

Zirin, H. and K. Tanaka, 1973. The flares of August. Solar Phys. **32**, 173-207.

Zirin, H., 1984. Evidence for magnetic field rearrangement in a solar flare. Astrophys. J. **281**, 884-885.

4. *Quantitative Prediction*

4.1 INTRODUCTION

Many researchers and some forecasters believe that access to quantitative, objective, numerical analysis would help the forecaster and improve the forecasts. To them it is clear that a formal summary of past experience would provide the forecaster with a useful tool and would remove much of the burden of weighing and integrating a mass of data, some of it contradictory and much of it redundant. To others, what is clear is that no formal system can ever incorporate all of the forecaster's experience and judgment.

This chapter describes some methods, achievements, problems, and possibilities of quantitative prediction. In the next section formal methods are described, with emphasis on two recent projects that compare numerical forecasts to actual forecasts. Section 4.3 deals with the character of solar data and the significant problems caused by inadequate data. Verification and evaluation of numerical forecasts are discussed in Section 4.4. Section 4.5 describes methods that have not yet been applied to flare prediction, including informal methods.

4.2 FORMAL METHODS

In concluding his report for the Forecasters Working Group at the Meudon Workshop in 1984, Chairman Joseph Hirman made a plea for translation of research results into quantitative form applicable to forecasting: "In addition to the attempt to understand the fundamental physics of STP phenomena, one of the roles of research is to develop objective techniques and models to improve the skill of the forecaster. It is important for researchers to make their research results usable in the form of numerical guidance for forecasters. Such numerical guidance can be based on theoretical models or derived from large historical data sets."

Short-term weather forecasts today are based partly on numerical guidance that includes a physical model. Equations of motion with observed values of temperature, pressure, and wind velocity describe wind and pressure systems and their evolution. For long-term weather forecasts, numerical guidance relies on statistical analysis of past data.

No predictive theoretical models of flares have yet been developed. A physical model for flares or active-region evolution may seem remote, but can be kept in mind as a goal. In the future, magnetohydrodynamic models may project magnetic structure and stress in a solar active region as circulation models now project pressure, temperature, and wind speed in Earth's atmosphere. For solar activity forecasting, statistical and historical guidance is a present possibility, limited mainly by lack of an adequate data base. Quantitative predictive methods based on statistics of past experience have been developed and tested, and will be described here.

Formal methods are characterized by mathematical rigor and are based on assumptions that the analyzed data have certain properties, such as normality. Both of the studies emphasized here are of this type. The methods adopted by Vecchia et al. (1980; VTCJ hereafter) are discriminant analysis and logistic regression. Neidig et al. (1981; NWSHF hereafter) used two versions of multivariant discriminant analysis (MVDA). In both studies the predicted variable was probability of occurrence of X-ray bursts and the predictor variables were selected from among those available to forecasters at NOAA's Space Environment Services Center (SESC). Data on predictor variables, on X-ray burst occurrence, and on their own forecasts were supplied by SESC. All in all, the two numerical studies are as comparable to each other and to the SESC forecasts as one could expect. Nevertheless, there remain significant differences that no doubt affect the results.

First, an informal comment on how formal numerical forecasting schemes seem to work. Given the need to predict flares and the belief that their occurrence may be influenced by a number of other observable events or conditions (predictors), what is the best way to use observed information about predictors to deduce information about the probable future occurrence of a flare? A formal method provides a formula that combines the information in all the predictors and estimates the probability of flare occurrence.

The user must define what would be helpful to know about future flares: an experimenter wants to catch a big one, a communicator wants to avoid even little ones. According to NWSHF "the . . . percentage of forecasts that are correct is the quantity of interest to a customer who cannot tolerate false alarms. A quite different requirement applies, however, in a situation where surprise flares are unwelcome." Thus, the first step is to define and measure the predicted value in a way that reflects its practical significance. A forecast of probability of occurrence of X-ray bursts in three size categories was chosen at SESC as best able to fill the needs of a variety of users.

The next step is examination of the relation of each predictor to the burst rate. Ideally, values of each predictor variable would be linearly related to corresponding values of the predictand, with scatter or variance independent of the type of predictor and independent of the magnitude of the predictor and predictand. For now, we'll note that if this is not true at the outset, it can sometimes be made more nearly true by substituting a function of the predictor variable (its square or its logarithm, for example). If a pair of predictors is so closely related that one determines the other, then the information supplied by one is redundant and can be omitted. There are various schemes to eliminate variables that contain only redundant information and to combine the remaining informative data. One such example is multiple regression analysis.

Typically, the analysis proceeds in steps. The predictor variable that leads to the greatest reduction of variance in the predictand is identified. A set of predictions based on this variable defines a new variance, and the predictor that most reduces this variance is the next variable selected and included in the next set of predictions. The process continues until adding a new predictor fails to produce a significant reduction of variance.

Alternatively, one can use all the possible predictor variables for a first approximation, then successively eliminate ineffective predictors, identifying at each step the predictor variable that provides the least reduction in variance and continuing until elimination of another predictor significantly increases the variance. Clearly, the definition of significance and the statistic chosen to measure variance will influence the result.

In practice, most methods select a predictor that

reduces the variance in one part of the range of predictand values, then in another part. This is seen in an analysis of formal flare prediction described by McLellan and Haurwitz (1967). The first variable selected as a predictor describes high activity and has a large coefficient for predicting high activity; the second and third selections describe moderate and low plage intensity and add information to the prediction of low flare rates.

A similar process can be followed in the results presented by Bartkowiak and Jakimiec (1986). They analyzed 21 variables as flare predictors. Their figure 3 shows "factor loadings," which describe how the variables are weighted. Successive vectors--groups of predictor variables--contain information about (1) the biggest bursts and flares, (2) moderately large bursts and flares, (3) small flares, (4) big bursts and flares on the next day, (5) moderately large bursts and flares on the next day, (6) magnetic-field configuration of the active region, and (7) type and area of sunspots in the active region. Bartkowiak and Jakimiec found considerable interdependence among the variables and determined that 75% to 90% of the information contained in the whole set of predictors can be represented by 8 or 9 principal components or factors.

A prediction scheme can be tailored to give maximum information about a chosen range of predictand values by "stretching" the scale for that range of values. This versatility means that it is worth putting some careful planning into defining needs, selecting a method, and defining and scaling the variables. In fact, the bulk of the effort of numerical analysis lies in preparing the data; additional analysis adds relatively little expense.

4.3 DATA

SESC forecasts the probability of occurrence of X-ray bursts in three size categories, labeled C, M, and X; peak flux in each category is 10 times greater than in the preceding category, and occurrence frequency is about 5 times smaller. The forecast estimates the probability that no burst will occur in the next 24-hour period (class N), or that the biggest burst will be of class C or greater, of class M or greater, or of class X. The subjective and numerical forecasts discussed here are made separately for each active region for each day and are called region-day forecasts.

The mathematical basis of many methods includes assumptions about the character of the variables: typically, that they are normally distributed, that the variance of the predictand is independent of its value, and that the predictor-predictand covariance is standard from one predictor variable to another. Examples of data that might fit these assumptions are errors in measuring a length, or test scores of a group of students. For such well-behaved variables, frequently occurring small deviations cluster symmetrically about the mean value; plots of values of each predictor against corresponding values of the predictand show scatter that varies little from plot to plot, or from one part of a plot to another.

The size distribution of solar flares, and of many of the variables used to predict flares, are quite different. Their distribution is better described by a power law with a negative index. Large events are very scarce, but as one considers smaller and smaller events, numbers increase dramatically. Tables 5 through 7 of VTCJ show that the magnitude of the standard deviation varies with burst magnitude and that for some variables, values of the standard deviation vary over a wide range. In an attempt to correct this condition, VTCJ rescaled a number of variables.

Although there are enormous discrepancies between the actual data and assumptions of normality, linearity, and uniformity of scale and of covariance that are the bases of some mathematical derivations, in practice these are less devastating than might be expected. This is because mean characteristics are used in decision-making stages of the analysis, and, for many statistics of a distribution, mean values are distributed normally, no matter how skewed the parent distribution may be (central-limit theorem). The larger the sample size, the less likely it is that peculiarities of the parent distribution will affect the results of the analysis.

Variable selection by a formal method does not always correspond perfectly to a preexisting perception of a "good predictor." For example, the variable MAGGRAD, the spatial gradient of magnetic field in the active region, appears to have many attributes of a good predictor: data are quantitative and continuous and its physical significance is clear. Yet it is outscored by MAGCLAS, a categorical variable that expresses magnetic complexity and includes a class, delta, that depends on high gradient. One quality MAGCLAS has that MAGGRAD lacks is completeness: 3739 observations compared to 1986. No doubt this is the reason for its success.

Those who have carried out quantitative prediction of

flares agree that the results should not depend strongly on the method. Rather, the most influential factor is the completeness, accuracy, and consistency of the input data. This point was made repeatedly and forcefully by VTCJ. A large part of their effort went into correcting obvious inconsistencies in the data. Difficulties with producing a reasonably error-free sample of continuing data prevented them from following up the training analysis with a verification analysis. They said, "The difficulties encountered in the course of the present study will preclude a rigorous and completely reliable analysis of information contained in the data base. Such basic problems, if not satisfactorily resolved, may result in future records which cannot possibly provide the statistical information of interest or importance" (pp. 5-6). They listed a number of specific recommendations for improvement in data-base management.

The problem with errors in the data base, although far from trivial, is presumably correctable. Less tractable is the problem of data that are incomplete because of clouds or instrument failure. NWSHF selected 15 of the 31 predictor parameters, then noted that only 510 of the 6095 region-day records contain all 15. They stated: "This is indeed a hardship for statistical analysis" (p. 12).

Another problem is that many of the flare-predictive parameters consist of subjectively assigned categories. NWSHF remarked: "There are several parameters (e.g., spot class, flare history, magnetic class) for which assigned values are based upon quantitative studies. Fortunately (or perhaps therefore!) these parameters are among those from which the objective forecast derives most of its skill" (p. 12).

Some of these data problems are unnecessary. Quantitative measures of X-ray burst flux values are available. They are translated into rough categories (C, M, X) at SESC, and the categories, rather than the continuous values, are used both as the predictand and as one of the predictors.

4.4 VERIFICATION AND EVALUATION OF THE OBJECTIVE STUDIES

Forecasts are verified by comparison to the outcome--in this case, the biggest burst that occurred in the forecast period. A natural way to evaluate probability forecasts would be to group together similar forecasts and to determine the distribution of outcomes within each group. Forecast skill would result in good agreement between the forecast probability and the fraction of positive outcomes. This method is, in fact, used at SESC (Gray and Slutz 1967). Their results show that populations defined by the forecast do indeed differ from one another. Gray and Slutz emphasized also the very small population of region days with big bursts. Forecast evaluation is discussed further in Chapter 6.

The NWSHF and VCTJ studies used somewhat differing methods of evaluating forecasts. NWSHF's numerical forecast lumped the big bursts together so the forecast is of the probability that the largest event is in one of three classes: smaller than C, equal to C, and bigger than C. The sum of these three probabilities is unity. SESC forecasts were cast into similar form by taking the largest burst category for which the forecast probability attained 0.5.

To determine coefficients in the numerical scheme, VTCJ dichotomized at each stage: for each region-day they made a forecast of N (no burst) or of C (meaning C, M, or X), then of N (meaning none or C) or of M (meaning M or X), etc. To evaluate and compare the forecasts, VTCJ used the Brier score, which is the mean value of the squared difference between forecast and outcome (skillful forecasts have a low Brier score). They presented scores for the SESC actual forecasts and for the two numerical methods, discriminant analysis (DA) and logistic regression (LR). Averaged over "event observed," the score for LR tends to be slightly better than the score for DA, and the SESC score worse than the other two. On the other hand, the average score is dominated by the days with no burst. If one looks at days when the biggest burst was C, the SESC forecasts are consistently better than either LR or DA. The sensitivity of the Brier score is illustrated in Chapter 6.

VTCJ omitted spotless regions, about one-third of the total, because spotless regions seldom have flares. Including successful forecasts of "no flare" from spotless regions would, no doubt, improve all the scores. The scores for LR and DA may be unrealistic because the same sample was used for determining the weights of the predictors (training) and for scoring. On the other hand, Vecchia (interview, 1984) considered that inconsistencies and errors in the data have a greater adverse effect on the numerical forecasts than on the actual forecasts. VTCJ concluded that LR is potentially useful for flare prediction.

Several different scores were presented by NWSHF; for example, fraction of forecasts that are correct and fraction of events of each type that are correctly forecast. One method, MVDA/CL (multivariate discriminant analysis with the Cooley and Lohnes procedure) consistently has a slightly better score than SESC. Analysis of a reduced sample with more complete predictor data yields scores that are higher, by a comparable amount, for all methods. This suggests again that improving the data is the surest route to improving the forecasts. NWSHF concluded that the MVDA/CL method shows skill that "becomes markedly evident when complete parameter representation is achieved in the data base" (p. 25).

Differences from one score to another are small compared to uncertainties in their interpretation. What can be concluded from the numerical studies is that they score well enough to promise significant aid to the forecaster in the task of integrating information from a variety of predictors.

Use of numerical guidance in meteorological forecasts was described by Hughes (1980). He evaluated Model Output Statistics (MOS) and station forecasts in terms of the skill score (the Brier score, normalized by a climatological forecast). Scores were compared for 5 stations for 3 periods: 1968-1971, 1972-1975, and 1976-1978. Hughes concluded that "the forecasters changed little, while the MOS scores continued to improve, narrowing the margin between them. MOS is certainly competitive now" (p. 60). Perhaps "helpful" should be substituted for "competitive," for numerical guidance is a tool for the forecaster. Hughes suggested that the best use that the forecaster can make of this tool is to prepare an independent forecast and then compare it to MOS guidance. Significant discrepancy calls for examination of the premises and data on which the differing forecasts are based.

4.5 INFORMAL METHODS, MAXIMUM ENTROPY METHOD

Development of artificial intelligence--AI--in computers offers an alternative to formal methods that may be worth exploring. An example of cluster analysis applied to flare forecasting was given by Burov et al. (1980). Using a computer and a numerical learning technique they arrived at the following rule, which, because of high current interest in artificial intelligence, we quote completely:

> within the coming 24 hours no flares of C-class or larger will occur in a given active region, if any of the following four conditions is satisfied:
>
> - no spots in the active region;
> - no kinks of the neutral line, no bright points,

and no isolated pole in region, no AFS present or new EFR emerges within existing spot group or within 5 degrees of it;

- region is located nearly 30 degrees of Hale boundary, no isolated pole in region, no AFS present or new EFR emerges within existing spot group or within 5 degrees of it;
- no bright points, no isolated pole in region, no AFS present or new EFR emerges within existing spot group or within 5 degrees of it, no dynamics in the group.

If none of these conditions is satisfied, a flare will occur, if any of the following conditions is satisfied:

- magnetic classification of beta, beta-gamma, or delta, and there are more than 3 kinks on the neutral line;
- the neutral line is oriented to east-west, hairpin, circular or reverse polarity region, and there are bright points along the neutral line or fluctuation;
- there are bright points along the neutral line or brightness fluctuation, and more than 3 kinks in the neutral line.

If in the active region there are no signs, that may be dashed with "sufficient" or "necessary", we may expect a flare only in the case, where within 24 hours at least one flare occurred in this region and the class of the group in Zurich classification modification exceeds 2 according to the table in (4) (Hirman and Flowers 1977).

The first four conditions express the idea that sunspots are necessary for a flare; in addition, it is necessary to have at least one precursor from each of the other three groups. The second paragraph lists sufficient conditions for flare occurrence, given the necessary conditions of the first paragraph. If these are not decisive, sunspot class and persistence are used as described in the third paragraph. This type of analysis may be closer than most numerical schemes to the processes of a human forecaster. Certainly, the application seems less mysterious than the machinations of the computer fed a somewhat variable diet of up to 32 solar predictors. The alternative groups of criteria appear to offer a means of dealing with missing data, but note that all of the chosen variables are among those for which data are most complete.

Groups in the Burov study may parallel the use of combination variables or Boolean combinations in MVDA. These are variables that are logical or mathematical combinations of predictor variables. In NWSHF's analysis, a combination variable called New No. 2 was the third most successful in reducing variance between predicted and actual values of the predictand. New No. 2 is the product of 4 different sunspot classifications. An example of a logical or Boolean combination is "sunspot area >500 or class delta or class gamma and no flares have occurred in region."

In another analysis, NWSHF introduced 20 additional combination variables, mainly squared values and rate-of-change of individual predictors. (VTCJ would call this a "transformation" or "rescaling" of the variables.) "As in the case of the original 6 [see Table 5.13], the additional parameters were derived on the basis of intuition" (NWSHF, p. 19). NWSHF concluded that "combination parameters, although their role is not fully understood, seem to improve forecast scores" (p. 25). VTCJ, however, warned against use of combination variables, unless they are constructed on the basis of physical reasoning: ". . . these represent interactions among variables and should be suggested by expert scientists. Rarely should functions be selected based on empirical evidence of association with flare occurrence."

Combination variables constructed for the regression analysis carried out by McLellan and Haurwitz (1967) were among the earliest selections. Critical analysis of these results, however, disclosed that the data were "overfit"; that is, the calculated values matched the actual values of this particular data set more closely than expected in view of accidental errors (M. Haurwitz, telephone interview, 1984). The overfitting would have been discovered in the course of the original analysis, had enough data been available for a verification run. These results, like those of the later VCTJ study, were limited by the labor of preparing a suitable data base.

In practice, combination variables appear to be so successful that there is a good deal of caution in accepting their values. An obvious characteristic is that they rigidly predefine the relative weights of the included predictors.

Rescaling of single variables, including powers and polynomials, exponential and logarithmic functions, seems to be legitimate and useful. One can imagine carrying out an analysis, the product of which is a combination of predictor variables, then introducing this particular combination as a combination variable in a new analysis, expecting to find no significant residual predictability in variables other than that combination variable. It might be a way to "retrain" a model to maintain effectiveness as conditions change with the sunspot cycle. The need to change the constants in the predictive formula as activity changes during the solar cycle has been discussed by Jakimiec and Wasiucionek (1980).

The cluster analysis developed by Burov et al. (1980) is rather like the informal intelligent systems described by Lenat (1984, p. 207):

> The primary source of power in these expert systems is informal reasoning based on extensive knowledge painstakingly culled from human experts. In most of the programs the knowledge is encoded in the form of hundreds of if-then rules of thumb, or heuristics. The rules constrain search by guiding the program's attention toward the most likely solutions. Moreover--and this distinguishes the heuristically guided programs from those relying on more formal methods--expert systems are able to explain all their inferences in terms a human being will accept. The explanation can be provided because decisions are based on rules taught by human experts rather than on the abstract rules of formal logic.

The approach of Burov et al. (1980) appears to have much interest: its similarity to "expert systems," its relative simplicity, and the possibility that it could offer an alternative guide for decisions when data are missing seem to make it worthy of further study, or at least of verification. Its simplicity would be a significant virtue in a practical interactive forecasting situation where the forecaster would need, on the one hand, to have enough trust of the system to be willing to use it, and, on the other hand, a healthy skepticism based on knowledge of its limitations.

A method that has recently found enthusiasts in many fields is the maximum entropy method (Skilling 1984). It is applied in engineering, image processing, and forecasting severe climatic conditions (Christensen 1981). Skilling said, "It claims to be applicable whenever one needs to estimate a single vector of proportions $\mathbf{p}$ (= p_1, p_2,..., p_n) from seriously incomplete data, which could be fitted equally well by many different vectors. . . . Maximum entropy always selects the simplest possible result, containing the bare minimum of structure needed to fit the data. Spurious detail is reduced as far as possible, which doubtless accounts for the practical success of the method." Skilling went on to describe recent mathematical justification of the method, based on consistency between different, legitimate ways of approaching analysis.

In the preface to the fourth volume of the sourcebook that he edited, Christensen (1981) described three principles that determine optimum proportions for the vector of dependent variables, partition of the independent variable space, and weight normalization. Weight normalization is determined by the principle of minimum local entropy exchange. Partition of predictand space is determined by the principle of minimum global entropy (or maximum global entropy exchange). Finally, the probability distribution is determined by the principle of maximum local entropy, or of minimum local entropy exchange. The terminology comes from thermodynamics via information theory, where important development of the method occurred. The maximum entropy principle is that entropy, the sum of the products of each predictor with its logarithm, takes its maximum value, subject to constraints of the data. (Note that this is the condition applied to flare-predictive data by VTCJ in their use of logistic regression.)

4.6 CONCLUSION

Among those with experience in objective forecasting, there is consensus on two points: (1) Almost all of the useful predictive information is contained in a set of 8 or 9 parameters; the other 75% of the data base must contain information that is either redundant or useless. (2) The parameter consistently identified as having predictive value is an index of the current level of flare activity (persistence).

VTCJ recommended that methods be developed in the direction of specific peculiarities of flare prediction. They concluded that "perhaps the single most important consideration for further investigations concerns the types of variables to be recorded and their scale and time of measurement. Understanding . . . may not be achieved if the informative variables are not first determined and then properly collected" (p. 35).

NWSHF concluded: "The MVDA/CL procedure may be capable of producing forecasts superior to any presently available using conventional, subjective techniques . . . we predict that with improvements in data consistency, as well as the inclusion of new, objective parameters in the future, the computer forecast scores will continue to improve."

These pilot studies suggest that solar forecasting could develop along a route parallel to weather forecasting, and that numerical guidance may prove to be a useful tool. Lack of an adequate data set seems to be all that prevents utilization of tested statistical methods. The possibility of a theoretical model that would project active-region physical conditions needs to be investigated. The informal methods of artificial intelligence, like that of Burov et al. (1980), are appealing and deserve a trial. One might expect, however, that a computer will be better at logistic regression than at imitating a forecaster, and that its first useful role in the forecaster-computer partnership may turn out to be storage and summary of a great deal of accurate information about past situations and their outcome.

4.7 REFERENCES

Bartkowiak, A., and M. Jakimiec, 1986. Relationships among characteristics describing solar active regions. Proc. of Meudon Solar-Terrestrial Predictions Workshop (1984), G. Heckman, M. Shea, and P. Simon (eds.), Boulder, Colo.

Burov, V., Hirman, J., Flowers, W., 1980. Forecasting flare activity by pattern recognition technique, in Solar Terrestrial Predictions Proceedings, vol. III, Solar activity predictions (R.F. Donnelly, ed.). U.S. Dept. Commerce, NOAA, ERL, Boulder, Colo. (Supt. Doc., U.S. Govt. Printing Off., Washington, D.C. 20402), pp. C-235-241.

Christensen, R., 1981. Preface to Entropy Minimax Sourcebook, v. 4. R. Christensen, ed. Entropy Limited, Lincoln, Mass.

Gray, T., and R. Slutz, 1967. A verification of the ESSA-Boulder Space Disturbance Forecast Center Forecasts. ESSA Tech. Memo. ERLTM-SDL11, Boulder, Colo.

Hirman, J., and W. Flowers, 1977. An objective approach to region analysis for flare forecasting. NOAA, SEL, Boulder, Colo.

Hughes, L.A., 1980. Probability forecasting--reasons, procedures, problems. NOAA TM NWS FCST 24, Silver Spring, Md.

Jakimiec, M., and J. Wasiucionek, 1980. Short-term flare predictions and their stationarity during the 11-year cycle, in Solar Terrestrial Predictions Proceedings, vol. III, Solar activity predictions (R.F. Donnelly, ed.). U.S. Dept. Commerce, NOAA, ERL, Boulder, Colo. (Supt. Doc., U.S. Govt. Printing Off., Washington, D.C. 20402), pp. C-54-63.

Lenat, D., 1984. Computer software for intelligent systems. Scientific American, 251, 204-213.

McLellan, C., and M. Haurwitz, 1967. Objective prediction of solar flares, solar proton events (PCE), and geomagnetic activity by the REEP program. ITSA Tech. Memo. no. 91, Boulder, Colo.

Neidig, D., P. Wiborg, P. Seagraves, J. Hirman, and W. Flowers, 1981. An objective method for forecasting solar flares. AFGL-TR-8-0026, Air Force Geophysics Lab., Hanscom AFB, Mass.

Skilling, J., 1984. The maximum entropy method. Nature **309**, 748-749.

Vecchia, D., P. Tryon, G. Caldwell, and R. Jones, 1980. Statistical methods for solar flare probability forecasting. Rept. AFGL-TR-0-0336, Air Force Geophysics Lab., Hanscom AFB, Mass.

5. *Operational Flare Forecasting*

5.1 INTRODUCTION

Timely prediction of flares is attempted at only a few forecast centers; at others, services are limited to forecasting flare effects after the flare itself has been observed. Although we are concerned here with forecasting the flare, we need to keep in mind that practical interest in flare prediction is based on the need to predict the consequences of flares, and that much of the forecasting effort goes into collecting and analyzing data on flare effects and predicting these effects.

Users' Needs

Users' needs determine the type of forecast that is made, when it is made, how it is expressed, and how it is distributed. At a workshop on Solar-Flare Prediction at Stanford University in February, 1985, Joseph Hirman, chief forecaster at SESC, identified important needs of flare-forecast users and described the types of forecast that meet these needs:

1. Users concerned with the effects of enhanced X-ray flux on the ionosphere and on space systems, are provided with SESC forecasts of peak flux of soft (1-8 Angstrom) X-ray bursts.
2. Users interested in the effects of solar radio bursts on telemetry and on tracking of space vehicles are alerted to high levels of flare activity by this same forecast.
3. Users concerned about the effects of solar protons

are warned upon observation of a flare likely to be associated with acceleration of energetic particles.

4. Users concerned with geomagnetic disturbance are advised when mass ejection is observed or deduced to accompany a solar flare.

Forecast users were classified in a different way at the Boulder workshop on Solar-Terrestrial Predictions in 1979: first, those who consider the forecast in deciding when to begin or continue an observational program; second, those who need warnings; and third, those who use the flare forecast to make their own forecasts of some other parameter. Examples of each type of use are cited in the Report of the Working Group on Short-Term Prediction (Simon et al. 1979).

```
HFUS 3 BOU 212200
FROM SPACE ENVIRONMENT SERVICES CENTER BOULDER COLORADO
SDF NUMBER 142
JOINT USAF/NOAA REPORT OF SOLAR AND GEOPHYSICAL ACTIVITY
ISSUED 2200Z 21 MAY 1984
8.  ANALYSIS OF SOLAR ACTIVE REGIONS AND ACTIVITY FROM 20/2100Z TO 21/2100Z:
SOLAR ACTIVITY HAS BEEN VERY HIGH THIS PERIOD.  THE LARGEST EVENT OF THE
PERIOD WAS AN X10/2B FLARE WHICH OCCURRED AT 2224 UT ON 20 MAY.  THIS EVENT
WAS ACCOMPANIED BY LARGE BURSTS THROUGHOUT THE RADIO SPECTRUM.  THESE INCLUDED
BURSTS AT 2695 AND 245 MHZ, OF 14000 AND 8500 FLUX UNITS, RESPECTIVELY.  A
TYPE IV RADIO SWEEP OF IMPORTANCE 2 WAS ALSO OBSERVED WITH THIS FLARE.
ANOTHER FLARE AN X2/2B OCCURRED ATD 2018 UT TODAY.  INITIAL REPORTS SHOW THAT
A TYPE II RADIO SWEEP WAS ASSOCIATED WITH THIS EVENT, NO OTHER REPORTS OF
RADIO BURSTS ARE AVAILABLE AT THIS TIME.  REGION 4492 (S10E34) WAS THE SOURCE
REGION FOR BOTH OF THESE FLARES.  THIS REGION HAS SHOWN GROWTH IN ITS TRAILER
PORTION, AND REMAINS A F TYPE SPOT GROUP WITH A BETA-GAMMA-DELTA MAGNETIC
CONFIGURATION.  REGION 4494 (S09E54) HAS PRODUCED SEVERAL SMALL FLARES AND IS
STABLE.  ONE NEW REGION WAS NUMBERED THIS PERIOD, 4495 (S07W43), A B TYPE SPOT
GROUP.
IB.  SOLAR ACTIVITY FORECAST: SOLAR ACTIVITY SHOULD BE MODERATE TO HIGH
THROUGHOUT THE FORECAST PERIOD.  REGIONS 4492 AND 4494 ARE THE MOST LIKELY
CANDIDATES FOR SIGNIFICANT FLARE ACTIVITY.
IIA.  GEOPHYSICAL ACTIVITY SUMMARY FROM 20/2100Z TO 21/2100Z: THE GEOMAGNETIC
FIELD HAS BEEN AT STORM LEVELS, AT ALL LATITUDES, THIS PERIOD.  THIS ACTIVITY
IS PROBABLY DUE TO RECURRENT CORONAL HOLE STREAMS.
IIB.  GEOPHYSICAL ACTIVITY FORECAST: THE GEOMAGNETIC FIELD SHOULD BE AT ACTIVE
TO STORM LEVELS, AT ALL LATITUDES, THROUGHOUT THE FORECAST PERIOD.
III.  EVENT PROBABILITIES 22 MAY - 24 MAY
CLASS M          90/90/90
CLASS X          40/40/40
PROTON           15/20/20
PCAF             RED
IV.  OTTAWA 10.7 CM FLUX
OBSERVED         21 MAY 140
PREDICTED     22-24 MAY 142/148/150
90-DAY MEAN      21 MAY 130
V.  GEOMAGNETIC A INDICES
OBSERVED AFR/AP            20 MAY 21/37
ESTIMATED AFR/AP           21 MAY 36/45
PREDICTED AFR/AP        22-24 MAY 40/40 - 30/30 - 20/30
SOLTERWARN
BT
```

Figure 5.1--An example of the daily "2200Z" forecast from SESC.

The Forecast

The forecast that is of most interest here is the daily probability forecast. Figure 5.1 is an example of the primary "2200 forecast" for an active situation, and Table 5.1 summarizes its contents. This example includes the forecast that is the focus of most of our interest: a quantitative estimate of probability of occurrence of X-ray bursts of each size for each of the next 3 days. It contains as well a verbal report of the solar conditions most influential in determining the flare forecast--magnetic fields in active regions and flare events that have occurred. Thus optical flares are reported, although X-ray flares are forecast. Quantitative general indices of activity, current and predicted, are the 10.7 cm wavelength radio flux from the Sun and the index of geomagnetic activity.

Table 5.2 is a list of SESC products provided by Hirman. It reemphasizes that flare prediction is only a small fraction of the total effort of gathering and distributing information about solar activity and its geophysical effects.

TABLE 5.1

CONTENTS OF SESC PRIMARY ROUTINE FORECAST: DAILY AT 2200 UT

- I. Solar activity description
  - A. Conditions and events observed in last 24 hours
  - B. Forecast for next 3 days
- II. Geomagnetic description
  - A. Conditions observed in last 24 hours
  - B. Forecast for next 3 days
- III. Flare forecast
  - A. X-ray burst, Class M, (probability)
  - B. X-ray burst, Class X (probability)
  - C. Polar-cap absorption (categorical)
- IV. Quantitative solar index: 10.7 cm radio flux
  - A. Observed in last 24 hours
  - B. Predicted for next 3 days
  - C. Predicted, 90-day mean
- V. Quantitative geomagnetic index: AFR, AP
  - A. Observed, yesterday
  - B. Estimated for today
  - C. Predicted for next 3 days

TABLE 5.2

SESC PRODUCTS

REAL-TIME

ALERTS
PRESTOS
STRATWARMS
SMM SUPPORT
GWC BACK-UP
SATELLITE BROADCAST MESSAGE
GWC DIRECT DATA LINK
NOSC DIRECT DATA LINK
SELDADS CALL-UP USER DISPLAYS

3 HOURLY

BOULDER K-INDEX
ANSAPHONE RECORDING OF SOLAR ACTIVITY
WWV RECORDING OF SOLAR ACTIVITY
NASA SHUTTLE SUPPORT

DAILY

SOLAR & GEOPHYSICAL ACTIVITY REPORT AND FORECAST (SGARF)
SGARF - PROBABILITIES ONLY
BOULDER RWC ADVICE
GEOALERT
SOLAR REGION SUMMARY
SOLAR AND GEOPHYSICAL ACTIVITY SUMMARY
1800Z FORECAST (Kp & preliminary 10cm Flux)
BOULDER SPOTS REPORT
BOULDER NEUTRAL LINE MAP
BOULDER CORONAL HOLE MESSAGE
DATA ACQUISITION REQUEST
ANCHORAGE ADVISORY REPORT 2100Z
SST RADIATION FORECAST
H-ALPHA PRINTS
DAILY RGON STACK PLOTS
Retransmission of ATN messages
- SMM Observing Schedule
- ANCHORAGE URALS DATA MESSAGE
- BOULDER UFOFH MESSAGE
- Fredericksburg UMAGE MESSAGE
- CULGOORA MORNING/AFTERNOON/EVENING REPORTS
- GEOALERTS - DARMSTADT, PARIS, MOSCOW, TOKYO, SYDNEY
- STANFORD SOLAR MEAN FIELD
- IUWDS DATA INTERCHANGE
- IUWDS QUALITY CONTROL LISTINGS

WEEKLY

PRELIMINARY REPORT OF SOLAR GEOPHYSICAL DATA
SYNOPTIC MAP
GEOSYNCHRONOUS SATELLITE ENVIRONMENT GOES-5
27 DAY AP & 10CM FORECAST
GEOALERT ADVICE SUMMARY - RWC'S

MONTHLY

MONTHLY SUMMARY OF GEOALERTS & PRESTOS FOR SGD
LISTINGS / MAPS / DRAWINGS FOR SGD/NESDIS
NAVAL RESEARCH LABORATORY DATA TAPE
NESDIS DATA LISTINGS
NESDIS DATA TAPE
USAF DATA LISTINGS
Retransmission of the
SIDC PROVISIONAL INTERNATIONAL MEAN MONTHLY SUNSPOT NUMBER

3 TO 12 MONTHS

IUWDS CODE BOOK
QUARTERLY MAGPRAG AND 7 DAY FORECAST VERIFICATION
FORECAST VERIFICATION
SUN-SATELLITE PROXIMITY AND ECLIPSE TIMES
SMOOTHED SUNSPOT NUMBER PREDICTIONS
NASA SHUTTLE SUPPORT SIMULATIONS

ARCHIVED & SENT ON REQUEST

REAL-TIME DATA & PRODUCT ARCHIVES
DAILY SOLAR INDICES; MONTHLY & YEARLY SUMMARIES
LIST OF M & X FLARES
LIST OF MAJOR EVENTS
GOES RAW ARCHIVE TAPES
TIROS RAW DATA ARCHIVE TAPES
NOVA2 1-5 MINUTE DATA BASE ARCHIVE TAPES
SELDADS PROCESSED DATA TAPES
MONTHLY SESC VERIFICATION

USER INFORMATION

SESC GLOSSARY OF SOLAR TERRESTRIAL TERMS
SESC PRODUCTS AND SERVICES BROCHURE
REPRINTS
SELDADS OPERATORS DESCRIPTION & USERS MANUAL

==

Forecast Distribution

Distribution of forecasts is also determined by user needs. SESC forecasts on demand, that is, the forecaster on duty responds to users who telephone to request information pertinent to a specific event or to a specific operation. Certain users are alerted when X-ray flux exceeds a defined threshold; other users are warned when thresholds are exceeded by radio flux, by energetic-proton flux, or by geomagnetic-disturbance index.

A world-wide distribution system for forecasts and data is provided by the International Ursigram and World Days Service (IUWDS), an "inconspicuous but effective arm of the nongovernmental scientific organizations" (Shapley 1973). This network includes Regional Warning Centers (RWC) at Boulder, Paris (Meudon), Darmstadt, Moscow, Tokyo, and Sydney, with another planned at Ottawa. Associated RWC are located in Stockholm, Prague, New Delhi, and Irkutsk.

Table 5.3 describes the six Regional Warning Centers. At Moscow and at Boulder 24-hour operations continue 7 days a week. Most of the centers send out a data message and a forecast daily; SESC in Boulder, which is designated also as the World Warning Agency, sends three data messages and a forecast each day. The information in Table 5.3 shows that the centers that participate in this international cooperation have primary concerns that vary widely, ranging from astronomical research to communication via the ionosphere.

Forecast Methods

Forecasting methods at different centers reflect this variety of concerns. Outstanding efforts in flare prediction have been made at l'Observatoire de Meudon (Paris RWC). The 1966 Proton Flare Project, under the auspices of IUWDS and the leadership of Paul Simon, successfully forecast in advance a number of proton flares, facilitating an intense observational effort focused on these most energetic events. The Paris RWC is closely tied to a vigorous research group, and to sources of observational data. The forecast is based on careful study of magnetograms and H-alpha and CaII K-line images. Chromospheric structure, its relation to the photospheric longitudinal magnetic field, and the evolution and spatial relations of dynamic features are stressed in the analysis and prognosis carried out by this group (Simon, 1973, 1979).

In the Soviet Union, three institutions have been

TABLE 5.3

REGIONAL WARNING CENTERS OF THE INTERNATIONAL URSIGRAM AND WORLD DAYS SERVICE

1) Space Environment Services Center, NOAA, BOULDER, Colorado, USA

Data: Solar and geophysical data for IUWDS interchange issued thrice daily.
Alert: PRESTO when flare emission observed.
Forecast: Regular solar and geophysical activity summary and forecast issued daily at 2200 UT.
Operating hours: continuous

2) Observatoire de Meudon, PARIS, France

Data: URSIgram: solar and geophysical data; summary of solar and geophysical data and disturbances.
Forecast: Geoalert: Solar activity and geophysical disturbances; recurrent geomagnetic activity.
Operating hours: 0730-1630 UT, Monday through Friday; 1300-1500, Saturday; closed Sunday and holidays.

3) Institute of Applied Geophysics, MOSCOW, USSR
Data: URSIgram: solar and geophysical data issued daily.
Forecast: Solar activity and geophysical disturbances; ionospheric disturbances in URSIgram for National Centers.
Operating hours: continuous.

4) Radio Research Laboratories, Ministry of Posts and Telecommunications, TOKYO, Japan
Data: URSIgram: solar and geophysical data issued daily.
Alert: PRESTO as appropriate.
Forecast: Solar activity and geophysical disturbance daily; 7-day radio telecommunications semiweekly.
Operating hours: 2330-1300 UT, 7 days/week.

5) Ionospheric Prediction Service, Darlinghurst (SYDNEY), Australia
Data: URSIgrams: solar and geophysical data at 0100 UT daily; occasionally at 0600 UT, and following major events.
Forecast: Daily in "Advice" message to World Warning Agency and occasionally as required to organizations mostly in Australia, New Zealand and Antarctica.
Operating hours: 2230-0630 UT and later as required. Sometimes closed from 0300 UT on Saturdays, Sundays, and holidays.

6) Forschungsinstitut der Deutsche Bundespost, DARMSTADT, GFR
Data: URSIgram: solar and geophysical data daily at 1100 UT, Saturdays, Sundays, and holidays excepted.
Operating hours: 0600-1430 UT, closed Saturdays, Sundays, and holidays.

involved in forecasting flares. The Institute of Applied Geophysics (IAG) in Moscow is the Eurasian Regional Warning Center of IUWDS. This is a full-time, real-time operation. In addition to geophysical forecasts and solar-proton extrapolations, it issues daily qualitative forecasts of flare activity for a two-day period and weekly forecasts for a 27-day period (Avdyushin et al. 1979). The Crimean Astrophysical Observatory pioneered in using measurements of vector magnetic fields and their gradients to forecast flares, in particular, proton flares during cosmonaut flights. At the Astronomical Observatory of the Kiev University in the 1970's, daily flare forecasts were based on characteristics of sunspot groups (Severny et al. 1979).

From Czechoslovakia, the Astronomical Institute in Ondrejov issues a weekly forecast of solar activity based on sunspot groups and facular fields, solar radio emission and ionospheric effects of flares (Krivsky 1981).

At the Toyokawa Observatory of the Research Institute of Atmospherics of Nagoya University in Japan, the main item of prediction in 1979 was the occurrence of proton flares (Enome 1979). This effort began in 1966 with the Proton Flare Project and has continued with other international campaigns. Enome noted the success of this effort: Among the occasions of positive prediction, the actual occurrence of proton flares was enhanced by a factor 20. These forecasts were based on active-region microwave flux at 3 cm and 8 cm wavelength, and on the spectral slope of the enhanced emission, measured by the ratio of fluxes at these two wavelengths.

In China, the Solar Activity Prediction Group of Peking Observatory (1979) predicts solar proton events and SID's 72 hours in advance, and also makes monthly and solar-cycle (11-year) predictions of solar activity. Chen and Zhao (1979), of the Purple Mountain Observatory in Nanking, reported at the Boulder Workshop on a quantitative method of analyzing data on sunspots and magnetic fields in active regions. The aim of the project was "to foresee the level of activity of every active region on the solar surface in the coming 1-3 days and thereby to estimate the possibility of the occurrence of ionospheric disturbances and proton events."

At Yunnan Observatory in China, Ding and Zhang (1979) studied the twisted pattern of penumbral filaments seen in well-resolved images of "spiral sunspots." From this twist they inferred large-scale rotation of sunspots and the magnetic field that leads to the presence of strong electric currents.

Observers make a practical prediction of the location of the next flare when they point a telescope with limited field of view at a given active region. This type of forecasting was an important part of the Skylab and Solar Maximum Missions and it presently goes on at Big Bear Solar Observatory and at the VLA.

Forecasters at Boulder study images that show magnetic field and chromospheric structure and also consider a variety of qualitative and quantitative data about active regions and the events that occur in them.

Collaboration between the Space Research Center of the Polish Academy of Sciences in Warsaw and the Astronomical Institute of the University of Wroclaw led to development of real-time numerical forecasts (Stasiewicz et al. 1979). Solar data are the input to another computer program that predicts radio propagation conditions. These are described in more detail in Section 5.5.

These examples show great variety in the methods and data bases used at different centers. This variety extends to the object of the forecast: at Meudon, for example, activity is described in four categories: "Quiet," "Eruptive," "Active," or "Proton Event," whereas at Boulder, the probability of X-ray bursts is forecast.

Many forecasting centers around the world do not forecast flares directly. They receive the flare forecast along with other data from the major centers and use it for their own purposes. For example, the U.S. Air Force Global Weather Central (GWC) uses forecasts and data from SESC as input to its own product: predictions of radio-wave propagation and other geophysical effects of the flare.

5.2 DATA ACQUISITION

This section describes in more detail how solar and geophysical data are gathered at the World Warning Agency and RWC at Boulder. The IUWDS (1973) publication "Synoptic Codes for Solar and Geophysical Data" lists stations contributing data; they cover the globe, from Resolute Bay to South Pole and from Ougadougou to Kislovodsk. Data are coded into 5-digit "words" for efficient telephone and telex transmission among key locations around the globe.

Data flow into the SESC communications hub from observatories in space and on the ground, providing information about a wide variety of solar emissions and conditions and about the state of the geomagnetic field and the ionosphere. The forecaster's console includes 13 computer screens, a digital counter, 3 teletypes, and several telephones with dedicated lines. An array of recent data is at arm's reach behind the forecaster. Forecasts, neutral-line drawings, and H-alpha images for the last month can be produced at a moment's notice. Those for the last year are also easily accessible. Code books hang within easy reach, computer commands are posted at the terminal, and quick-dial codes are posted at the outgoing telephone. The history and current values of 5 basic parameters are on display. Close by, other teletypes feed information to the computer. One of the functions of the computer is to monitor the incoming data stream and alert the forecaster to any of 20 situations that might need immediate attention.

SESC is operated jointly by the National Oceanic and Atmospheric Administration (NOAA) of the U.S. Department of Commerce and the Air Weather Service of the U.S. Department of Defense. SESC and GWC are separate centers, each serving a specific group of users. Coordination of efforts permits sharing of data, communication links, personnel, and some responsibilities for services and products.

Three basic sources of solar-flare data are the SOON optical network (Holloman 1982), the RSTN radio observatories, and the GOES X-ray monitor. SOON and RSTN are operated by the Air Weather Service; GOES is a NOAA satellite. The aim of SOON is to observe and report solar data in real time, continuously and automatically, using standardized equipment and procedures. The system, designed by Richard Dunn of Sacramento Peak Observatory, includes a high-resolution telescope, a tunable birefringent filter, a videometer, and a computer. The result is large-scale on-band and off-band H-alpha images and magnetograms of selected active regions, available immediately in digital form, and eventually on film. One way of presenting the digital data on H-alpha intensities is as a histogram of pixel brightness (Fig. 5.2). Automatically a flare is defined when a certain number of pixels exceed a given threshold of brightness. Ideally, observers at the various sites calibrate and coordinate pointing of the telescopes at the various active regions on the Sun's disk, specify the background brightness in each region, and define flare thresholds. If this is accomplished, and the filters are similarly tuned, the result should be an automatic measurement of flare area and brightness as a function of time, providing a quantitative and standardized description of the H-alpha flare.

Magnetometer measurements are displayed as spatial contours of the line-of-sight magnetic field (Fig. 5.3), and as contours of the gradient of this field (Fig. 5.4). The

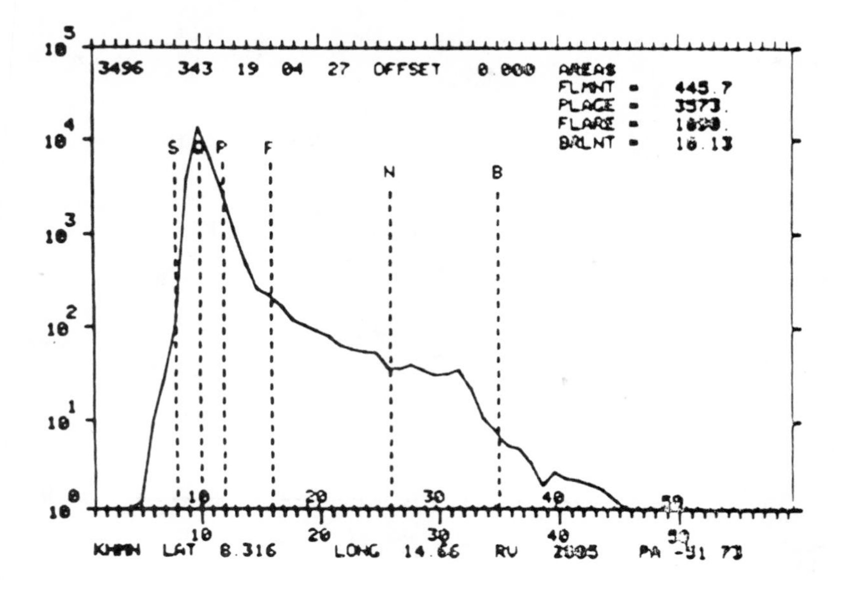

Figure 5.2--An example of a SOON-site histogram of pixel brightness.

system can analyze H-alpha, white-light, and magnetometer data from six active regions with 5-second time resolution. The RSTN sites detect radio bursts at eight frequencies (wavelengths from 3.4 cm to 1.2 m). The shape of the radio spectrum gives an early indication of emission of energetic charged particles.

The GOES satellites provide information on X-ray bursts in the spectral ranges 1-8A and 0.5-4A. Time sequences of X-ray flux are shown in Figure 2.13. These data are basic for SESC flare forecasting because X-ray peak flux is the quantity forecast. The GOES satellites also detect energetic-particle enhancements. Along with reports of H-alpha flares from one or two additional observatories, the SOON, RSTN, and GOES data provide most of the real-time information on flares at SESC. An important facility is the capability of questioning SOON and RSTN observers to confirm reports of unusual activity. These near real-time data are handled well, according to Hirman. Older data take more overhead, he explains: If you cannot get back to the original observer, you cannot confirm a puzzling piece of information without a lot of expended time. Data that arrive through the international exchange once a day (or sometimes

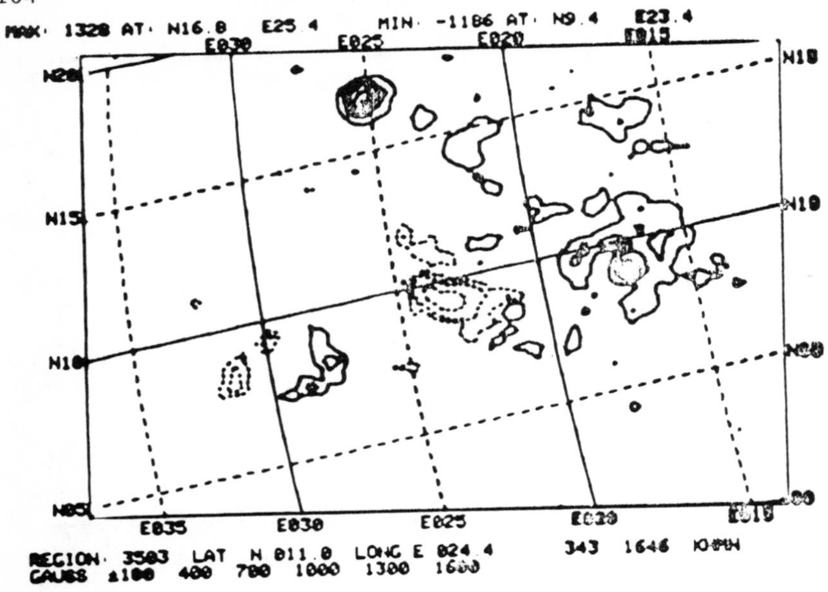

Figure 5.3--A SOON-site magnetogram of the longitudinal magnetic field.

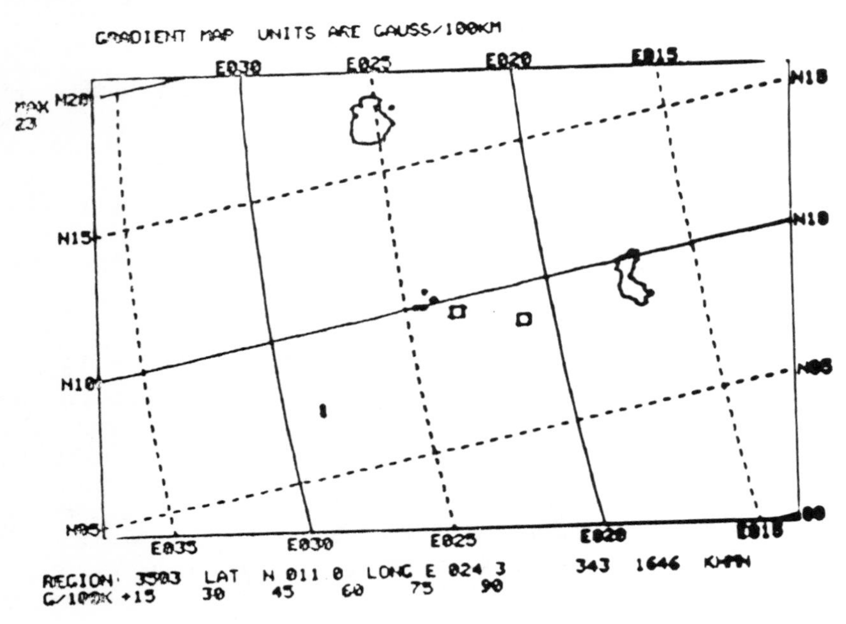

Figure 5.4--A SOON-site plot of gradient of the longitudinal magnetic field.

only for special events) are valuable in hindsight but not for "now-casting."

These valuable near real-time data, listed in Table 5.4, include digitized fluxes and indices, coded reports, verbal descriptions, and images, drawings, and graphs. A computer scans the incoming data for obvious anomalies, and aids further by producing lists and summaries on demand and by calculating future fluxes expected on the basis of current activity.

About 800 sets of solar and geophysical data are available through the Space Environment Laboratory Data Acquisition and Display System (SELDADS). Fig. 5.5 shows how data flow from ground and space observatories to the user, to the forecaster, and to the archives. Table 5.5 lists types of data that make up the product of this computerized data base, still under development. It gives some idea of the enormous amount and variety of information that the forecaster must absorb and transform into alerts and warnings, forecasts, and outgoing reports of relevant facts. The forecaster is immersed in a flood of evolving information and must quickly assimilate and evaluate these data and continually fine-tune his forecast. Hirman's illustration of the process appears as Figure 5.6.

The forecaster is sometimes overwhelmed and saturated with data; on other occasions poor weather conditions may result in a dearth of information; nevertheless, the forecast must be made. Would it help to include with the forecast of probability of flare occurrence an indication of the forecaster's confidence in the forecast and the data on which it is based? This could be done by expressing the forecast as a wide or narrow range of probabilities.

Data collected at SESC and entered into the computer data base are retained for 30 days. Archiving and publishing solar and geophysical data is the responsibility of the National Geophysical Data Center of the National Environmental Satellite Data and Information Service (NESDIS). Although SESC and NGDC are both NOAA groups and both are in Boulder, they are separate entities and deal with different data, acquired by different means. The real-time information that flows through SESC is not necessarily inferior to the archived data; according to Hirman, "we try harder," especially for the big events. In any case, after 30 days no record remains of the exact data on which the forecast was based. This is in keeping with the real-time focus of forecasters' concern: there is little time in their day (or night) for reflection or retrospective assessment of the task.

TABLE 5.4

NEAR "REAL-TIME" DATA AVAILABLE AT SESC FORECAST CENTER

	FREQUENCY of REPORT	PERCENTAGE of time AVAILABLE
CODED DATA:		
Solar optical		
Sunspot	5/day	90%
Flare	activity dependent	98%
Region analysis	5/day	10%
Features	5/day	90%
Histograms	1/hour	90%
Indices	5/day	90%
Solar radio		
Events	activity dependent	99%
Indices	5/day	95%
Ionospheric		
IEC	1/hour	98%
Ionosonde	1/hour	98%
Auroral radar	1/15 min	95%
Geomagnetic		
Component	1/min	98%
Events	activity dependent	98%
Indices	1/three hours	98%
COMPUTER GENERATED PRODUCTS:		
Models		
Probabilities	on demand	98%
Proton	on demand	98%
10cm	on demand	98%
Numerical Guidance	on demand	98%
Activity Summaries	on demand	98%
Listings	on demand	98%
Events	on demand	98%
Regions	on demand	98%
Magnetic	on demand	98%

	FREQUENCY of REPORT	PERCENTAGE of time AVAILABLE
RAW DATA:		
Images		
H-Alpha	2/day	80%
Calcium	1/day	80%
White light	1/day	75%
Maps		
Magnetic	3/day	80%
Green line (5303A)	1/day	55%
Yellow line (5694A)	1/day	55%
Neutral line	1/day	90%
Helium (10830A)	1/day	80%
Drawings		
Sunspot	1/day	80%
E-W radio scan	1/day	90%
Spacecraft Data		
Solar xray	1/min	99%
Magnetic	1/min	99%
Particle	1/five min	99%
Interplanetary	1/five min	33%
DESCRIPTIVE DATA:		
Plain language		
Observatory	8/day	95%
Special events	activity dependent	98%
Forecasts		
Primary	1/day	99%
HF propagation	4/day	99%
GEOALERT advices	6/day	88%
WEEKLY	1/week	99%

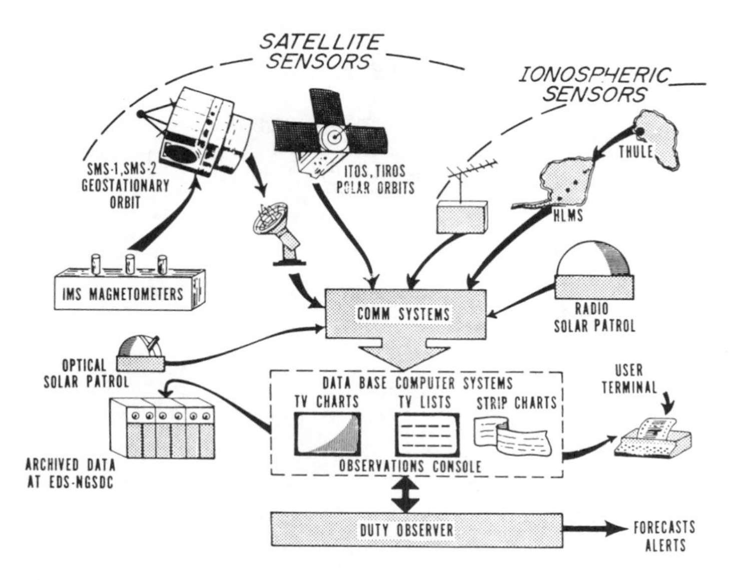

Figure 5.5--SEL real-time data acquisition and display at SESC (SELDADS).

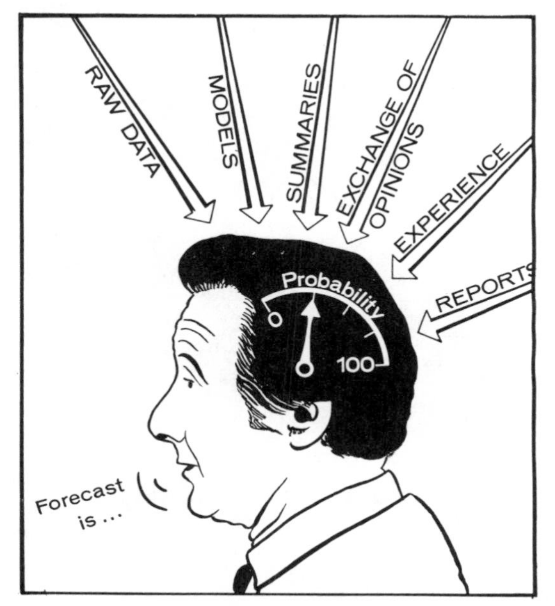

Figure 5.6--The mental process of forecasting. (Courtesy of Joseph Hirman, SESC.)

TABLE 5.5

SELDADS I DATA SETS

	SOLAR XRAYS AND ENERGETIC PARTICLES	
GOES XRAYS	1&5 MIN AVERAGES	GOES 5&6
GOES XRAY EVENTS	RT	GOES 5&6
GOES PARTICLE COUNTS	1&5 MIN AVERAGES	GOES 5&6
GOES PARTICLE FLUXES	1&5 MIN AVERAGES	GOES 5&6
GOES PARTICLE EVENTS	RT	GOES 5&6
XRAY BACKGROUND	RT & DAILY VALUE	GOES 5&6
PROTON FLUENCE (>1, >10 MeV)	DAILY	GOES 5&6
TIROS/NOAA PARTICLES		NOAA 7&8
SST RADIATION DOSE RATE		
RIOMETER	15 MIN	4 STATIONS
	1&5 MIN	5 STATIONS
NEUTRON MONITOR	15 MIN	THULE
ABSORPTION (UABSE)	DAILY	IUWDS
METEOR ENERGETIC PARTICLES	(USSR)	
GROUND LEVEL EVENTS - COSMIC RAYS		THULE NEUTRON MONITOR
FORBUSH DECREASES - COSMIC RAYS		THULE NEUTRON MONITOR
COSMIC RAY DAILY AVERAGE		THULE NEUTRON MONITOR
POLAR CAP ABSORPTION EVENTS		THULE RIOMETER
DAILY EVENTS EDITED - X-RAYS, PROTON, PCA, GLE, FORBUSH DECREASE, SST DOSE RATE		
EVENTS ASSOCIATED TO REGIONS - X-RAYS, PROTON, PCA		
INTEGRATED X-RAY FLUX FOR EVENTS		
	SOLAR OPTICAL EVENTS & REPORTS	
H-ALPHA FLARES	RT & DAILY	6 STATIONS + IUWDS
H-ALPHA LIMB/DISK EVENTS	RT	6 STATIONS
H-ALPHA PLAIN LANGUAGE REPORTS	RT	4 STATIONS
H-ALPHA PATROL TIMES DAILY	RT	6 STATIONS
H-ALPHA REGION ANALYSIS	DAILY	4 STATIONS
H-ALPHA HISTOGRAMS	DAILY	4 STATIONS
H-ALPHA PLAGE	DAILY	IUWDS STATIONS
CALCIUM REPORTS	DAILY	MANILA
SUNSPOT REPORTS	DAILY	5 STATIONS + IUWDS
OPTICAL WHITE LIGHT SPOTS		
SUNSPOT NUMBER	DAILY	BOULDER
MAGNETIC FIELD STRENGTH	DAILY	MT WILSON
MAGNETIC DATA		
MAGNETIC MAPS		SOON NETWORK
CORONAL HOLE		BOULDER
CORONAL LINE	DAILY	IUWDS STATIONS
EPHEMERIS		
RETURNING REGION		
DAILY SESC CONSENSUS REGION SUMMARIES		
DAILY EVENT EDITED - FLARES, LIMB/DISK		
EVENTS ASSOCIATED TO REGIONS - FLARES, LIMB/DISK		
SRI PROBABILITIES		
SESC DAILY SUNSPOT NUMBER		
FLARES BY REGION		
NUMBER OF NEW SUNSPOT REGIONS ON THE DISK		
	SOLAR RADIO EVENTS & INDICES	
SOLAR RADIO DISCRETE BURSTS		5 STATIONS + IUWDS
SOLAR RADIO SWEEP FREQ BURSTS		5 STATIONS + IUWDS
SOLAR RADIO NOISE		2 STATIONS + IUWDS
10 CM FLUX	14,17,20Z	OTTAWA
SOLAR RADIO BACKGROUND FLUX	DAILY	4 STATIONS + IUWDS
SOLAR MEAN FIELD	DAILY	STANFORD, THULE, VOSTOK
SOLAR RADIO PATROL TIMES	DAILY	3 STATIONS
DAILY EVENTS EDITED - RADIO BURSTS (DISCRETE & SWEEP), RADIO NOISE STORMS		
EVENTS ASSOCIATED TO REGIONS - RADIO BURSTS, NOISE STORMS		
INTEGRATED BURST FLUX FOR EVENTS		

Table 5.5 (continued)

	GEOMAGNETIC FIELD	
SATELLITE MAGNETIC FIELD	1&5 MIN AVE 3 COMPONENTS	GOES 5&6
HEMISPHERICAL POWER INPUT	90 MIN VALUES	NOAA 7&8 (TED)
EQUATORWARD BOUNDARY	90 MIN VALUES	NOAA 7&8 (TED)
Q-INDEX	90 MIN VALUES	NOAA 7&8 (TED)
H,D & Z VALUES	1&5 MIN AVE	12 RGON STATIONS
H, D VALUES	15 MIN DATA	THULE & ANCHORAGE
X,Y,Z MIN/MAX, K VALUES	90 MIN	AF HIGH LATITUDE NETWORK 7 STATIONS
AP, KP (running & daily)	3 HOURLY	AF HIGH LATITUDE NETWORK
H,D & Z	1 MIN AVE	USGS MAGNETOMETER
TOTAL FIELD	1 MIN	USGS MAGNETOMETER
K VALUES	3 HOURLY	USGS MAGNETOMETER
QUIET DAY CURVES	DAILY	USGS MAGNETOMETER
AFR & KFR VALUES	DAILY & EVENTS	FREDERICKSBURG
EVENTS - SSC, STORM BEGIN & END TIMES --		FREDERICKSBURG & AP
A & K VALUES	DAILY	20 IUWDS STATIONS
POLAR GEOMAGNETIC HOURLY VALUES	HOURLY	
INFERRED INTERPLANETARY FIELD DIRECTION		
IPS SOLAR WIND		
ISEE DATA		
ISEE SHOCK DETECTOR		ISEE
RGON SHOCK DETECTOR		RGON NETWORK
	IONOSPHERIC DATA	
SUDDEN IONO. DIS.	EVENTS	IUWDS STATIONS
TOTAL ELECTRON CONTENT	HOURLY	5 STATIONS
FOF2	HOURLY	IUWDS STATIONS
M3000	HOURLY	IUWDS STATIONS
FOES	HOURLY	IUWDS STATIONS
FMIN	HOURLY	IUWDS STATIONS
HF SENSOR DATA	15 MIN	ALASKA STATIONS
AURORAL RADAR	15 MIN	ANCHORAGE
CRITICAL FREQUENCIES	HOURLY	IUWDS STATIONS
IONOSPHERIC DATA	HOURLY	IUWDS STATIONS
OBLIQUE IONOSPHERIC SOUNDER		
	MISCELLANEOUS	
EQUIPMENT STATUS		
	ALERTS	
ALERTS		
STRATWARMS		
PRESTOS		
GEOALERT forecasts		
GEOALERT advice from RWC's		
	3-DAY FORECASTS	
M-CLASS XRAY EVENTS		
X-CLASS XRAY EVENTS		
PROTON EVENTS		
PCA		
OTTAWA 1700Z 10.7CM RADIO FLUX		
FREDERICKSBURG A-INDEX		
PLANETARY AP-INDEX		
NORTH PACIFIC PROPAGATION (Anch advisory)		
Flare probabilities by region		

5.3 FORECAST CRITERIA

As described in the introduction to this chapter, each forecasting group has its own method and primary predictor, from transverse magnetic field at the Crimean Astrophysical Observatory (and now at Marshall Space Flight Center), to radio flux and spectrum at Toyokawa. In this section we present lists of criteria used at forecast centers and in numerical predictions. The lists are long and varied. We'll try to impose some order on this situation by grouping together criteria that seem to be different ways of describing the same characteristic.

SESC's list (Table 5.6), compiled by director Gary Heckman (1979), is ordered according to the rank and weight that individual forecasters assign in considering various criteria and deciding on a forecast. Heckman noted that 56% of the forecasters placed persistence at the head of the list. Magnetic gradient, shear, spot class, and emerging

TABLE 5.6

SESC FORECASTERS' RANK OF FLARE PREDICTORS (Heckman 1980)

	Rank Range	Relative Weight (%)
High or rapidly increasing magnetic-field strength and gradient	1-7	11
Magnetic complexity	2-8	11
Persistence	1-12	11
Presence of shearing motion or proper motion indicating shear or twisted fields	1-10	10
Sunspot classification	1-8	10
Minor flares that are X-ray rich or high X-ray background	4-10	9
Emergence of new fields in stable or decaying portion of active region	1-10	7
Rapid spot growth after initial appearance	4-12	6
Colliding regions	4-12	6
Similarity of region to previous flare producers	8-14	6
Bright plage (or [high] McMath intensity)	3-13	3
Plage compactness	3-13	3

flux each got a vote for first place. Primary and Secondary Forecast criteria, described as traditionally used at SESC, are listed in Tables 5.7 and 5.8 (Heckman 1979).

The criteria used in active-region selection at Big Bear Observatory include (H. Zirin, written communication, 1984 Dec. 26): (1) magnetic structure, especially shear; (2) emerging flux producing field collision; (3) reverse polarity and delta configuration; and (4) filament activation. Zirin (1983) described flare precursors in one active region: "We were following [another region] which had sheared fields, and was tilted as well, when the occurrence of numerous small flares in 17760 led us to change our views on flare prediction and turn to it, relying on persistence."

TABLE 5.7

TRADITIONAL PRIMARY FORECAST CRITERIA AT SESC
(Heckman 1979)

Magnetic Field Configuration:

- Emergence, or growth, of new fields within, or near, existing fields such that the two fields interact.
- Merging of active regions through growth or proper motion.
- Polarities in bipolar region reversed from normal east-west order for hemisphere and solar cycle.
- Number and quality of convolutions in the radial neutral line.
- Magnitude and time-rate-of-change of gradients in radial component;
- Magnitudes of field strength maxima.

Subjective Reference to Past Activity:

- H-alpha plage intensity, structure, and size compared with past examples with known associated flare activity.
- Sunspot-group form and area compared with past examples with known associated flare activity.
- Region location with respect to active heliographic longitudes (see Dodson and Hedeman, 1972, and Svestka, 1972).
- Rate and manner of growth or decay.
- Flare activity during past few days, with greater consideration for the past 24 hours. Evolution of the region during the past 24 hours determines whether or not the observed level of flare activity will persist.

With persistence added to the list, the Big Bear set of predictors has much in common with the set used at SESC and those selected by numerical analysis.

Other lists are the criteria described by Severny et al. (1979) as used in the Soviet Union (Table 5.9), and predictor variables (Tables 5.10–5.12) used in numerical predictive efforts, which are described in Chapter 4. When availability and quality of data are a problem, good predictors are those for which data are quantitative and complete. Perhaps this is why, in speaking of numerical methods, Severny noted that "the most informative parameters . . . cannot be essential in the physical process of the flare."

TABLE 5.8

TRADITIONAL SECONDARY FORECAST CRITERIA AT SESC
(Heckman 1979)

Radio Emissions: Rise in background flux at any of several discrete frequencies (preflare heating)
X-ray Background Rise
Optical Coronal Enhancements (green and yellow lines)
Optical Activity at the Solar Limb
- Limb flares
- Coronal rain
- Surges, sprays
- Elevated plage (cap prominence)

TABLE 5.9

FLARE PREDICTORS USED IN THE SOVIET UNION
(Severny et al. 1979)

1. Horizontal gradient of the longitudinal magnetic field
2. Anomalies of the transverse magnetic field
3. Neutral line, parallel to the solar equator
4. The existence of moustaches and surges
5. The plage brightness in H-alpha
6. The relative brightness of H-alpha and CaII plages
7. Interacting plages (single 9-cm source)
8. Vertical gradient of magnetic-field strength
9. Rotation of sunspot group

TABLE 5.10

FLARE PREDICTORS IN PATTERN RECOGNITION STUDY*
(Burov et al. 1980)

Number	Parameter
1.	Spot Class (C)
2.	Magnetic Class (M)
3.	Magnetic field strength (H)
4.	Magnetic gradients (G)
5.	Sunspot dynamics (D)
6.	Stage of development (S)
7.	Leader emerged (L)
8.	Relationships with nearest sector boundary (B)
9.	Plage compactness (K)
10.	Main NL orientation (O)
11.	Neutral line complexity (N)
12.	Neutral line temporal changes (N)
13.	Associated filament (F)
14.	Bright points (P)
15.	Emerging flux (E)
16.	Largest flare (L)

* This study is described in Chapter 4.

In Bartkowiak and Jakimiec's (1986) study (Table 5.11) the neutral-line index, NL, combines information about orientation and complexity of the longitudinal field in the active region. Similar information is contained in parameters 11 and 12 of Table 5.12. The magnetic-field index, MFI, is the product of NL and the magnetic-field gradient. The predictors in Table 5.11 pertain to ongoing studies that continue work described by Jakimiec and Wasiucionek (1980). These studies are the bases of numerical prediction of solar activity and of radio propagation conditions. Numerical forecasting as routinely practiced at Warsaw is described in Section 5.5.

The variables in Table 5.12 are those used by Hirman et al. (1980) in numerical analysis of data for the year 1977. There are two extensions of this study. Vecchia et al. (1980) applied various methods of analysis to data for 1977, 1978, and January of 1979. Neidig et al. (1981) used two other procedures to analyze data from the same extended period. These studies are described in Chapter 4. In these

TABLE 5.11

FLARE PREDICTORS IN NUMERICAL STUDY
(Bartowiak and Jakimiec 1986, p. 2-3)

1. McI - Sunspot class of McIntosh--the variable is a product of qualitative representations of three McIntosh parameters.
2. A - Sunspot group area.
3. CaA - CaII plage area.
4. CaI - CaII plage intensity.
5. M - Magnetic class.
6. H - Strength of the magnetic field.
7. NL - Neutral Line index - the variable characterizing the orientation and complexity of the neutral line.
8. MFI - Magnetic Field Index - the variable is a product of the values of the magnetic gradient and the Neutral Line index. Full description of NL and MFI may be found by Jakimiec and Wasiucionek (1980)
9. C - The number of flares of C class per day.
10. MX - The number of flares of M class plus 10 times the number of flares of X class (per day).
11. Fs - Total flare flux in wavelength interval 1-8 Angstroms it is the sum of maximal values of X-ray flux of individual flares, which are observed in a sunspot group on a given day. It is an alternative description of the X-ray flare activity and can be used instead of the two separate variables: 9 and 10.
12. Fh - Total flare flux in wavelength interval 0.5-4 Angstroms.
13. sf - The number of faint subflares per day.
14. snb - The number of normal and brilliant subflares and flares of Imp 1f.
15. Imp - The number of flares of Imp 1n, Imp 1b, Imp 2f plus 10 times the number of flares of importance greater than 2f.

The eight following variables concern the flare activity observed on the next day:

16. C - The number of flares of C class per day.
17. MX - The number of flares of M class plus 10 times the number of flares of X class per day.
18. Fs - Total flare flux at 1-8 Angstroms.
19. Fh - Total flare flux at 0.5-4 Angstroms.
20. snb - The number of normal and brilliant subflares and flares of Imp 1f.
21. Imp - The number of flares of Imp 1n, Imp 1b, Imp 2f plus 10 times the number of flares of importance greater than 2f.

TABLE 5.12

PARAMETERS IN THE SESC REGION ANALYSIS PROGRAM

Parameter	Characteristics	Priority Scale
1. Spot Class	None observed	0
	Enter spot class three letter code	---
	No data	9
2. Magnetic Class	No spots	0
	Alpha	1
	Beta	2
	Beta-gamma	3
	Gamma	4
	Delta	5
	No data	9
3. Magnetic Field Strengths (largest)	No spots	0
	Enter letter (R/V) and two digit value (if same use leader polarity)	---
	No data	9
4. Magnetic Gradients in gamma/km	No spots or unipolar region	0
	Enter the gradient as N.NN	-.--
	No data	9
5. Sunspot Dynamics	No spots or not applicable	0
	Coalescing of spots	1
	Spot rotation	2
	Relative spot motion (opposite polarity spots)	3
	No data	9
6. Interaction with Another Region	None occurred	0
	Strong spots of opposite polarity converge (from less than 2 degrees apart)	1
	No data	9
7. Stage of Development	No spots	0
	Mature group (stable)	1
	Decaying	2
	Growing	3
	Rapid decay (spot or area decrease by > 50%)	4
	Rapid growth (spot or area increase by > 50%)	5
	Rapid growth (spot or area increase by > 100%)	6
	No data	9
8. Leader Emerged in Leader or Trailer Polarity Fields (from previous synoptic map)	Structure not definite	0
	Returning region - enter region number	---
	< 5 deg of NL and out-of-phase with NL	2
	> 5 deg of NL and in leader polarity fields	3
	> 5 deg of NL and in trailer polarity fields	4
	< 5 deg of NL and in-phase with NL	5
	No data	9

Table 5.12 (continued)

Parameter	Category	Code
9. Relationship with Nearest Sector Boundary (Hale = region polarity matches the boundary)	Sector structure not definite	0
	Region is > 30 degrees from nearest boundary .	1
	Non-Hale and 10 to 30 deg west of boundary ...	2
	Non-Hale and 10 to 30 deg east of boundary ...	3
	Non-Hale and < 10 deg of boundary	4
	Hale and 10 to 30 deg west of boundary	5
	Hale and 10 to 30 deg east of boundary	6
	Hale and < 10 deg of boundary	7
	No data	9
10. Plage Compactness and Embedded Filament (compact = NL corridor > 2 degrees wide)	Non-compact plage and no filament	0
	Non-compact plage with filament	1
	Non-compact plage with active filament	2
	Compact plage without embedded filament	3
	Compact plage with embedded filament	4
	Compact plage with active embedded filament ..	5
	No data	9
11. Main NL Orientation within Plage	Weak structure	0
	North-south (+/- 45 degrees to NS)	1
	East-west	2
	Hairpin (E-W)	3
	Mostly circular	4
	Reverse polarity region	5
	No data	9
12. Neutral Line Complexity	No kinks or weak structure	0
	1-3 kinks (very simple region)	1
	4-6 kinks (simple region)	2
	7-12 kinks (intermediate region)	3
	> 12 kinks (very complex)	4
	No data	9
13. Neutral Line Temporal Changes	No definite trend	0
	Neutral line becoming simple	1
	Neutral line becoming complex	2
	No data	9
14. Associated Filament (external to region but along common neutral line)	No associated filament	0
	Filament unchanged	1
	Filament growing	2
	Filament disappeared within past 24 hours	3
	Filament darkens or is active	4
	No data	9
15. Bright Points and/or Plage Fluctuations	None occurred	0
	Occurred but not along neutral line	1
	Occurred along the neutral line	2
	Plage fluctuations	3
	No data	9
16. Emerging Flux and/or AFS	None occurred or region is new	0
	Isolated pole in region	1
	New EFR emerges within existing spot group ...	2
	New EFR emerges near region (within 5 degrees of existing spot group)	3
	AFS present in region	4
	No data	9

Table 5.12 (continued)

17. Radio Burst and/or sweep	None occurred or small events	0
	> 250 flux units at 10 cm	1
	> 1000 flux units at 10 cm	2
	Type III sweep	3
	Type IV sweep	4
	Type II followed by type IV sweep	5
	U burst	6
	Major and complex 10 cm burst	7
	No data	9
18. Largest Flare Since Region Appeared	None occurred or first day observed	0
	C class flares have occurred	1
	M class flares have occurred	2
	X class flares have occurred	3
	No data or region appeared on east limb	9
19. Region First Appeared	Formed on disk	0
	Came around east limb - first transit	1
	Second transit	2
	Third transit (and etc)	3
	No data	9
20. Proton Event (this transit)	No particle event	0
	Proton-10 event = 10 $cm^{-2}sec^{-1}ster^{-1}$ @ >10 MeV	1
	Ground level event	2
	No data	9
21. (conventional forecast)		
22. Largest Flare for the Past 24 Hours	None occurred or < C	0
	Class C	1
	Class M	2
	Class X	3
	Proton event	4
	No data	9
23. 10 cm	10 cm flux value for today	---

two studies the selection, definition, and scaling of predictor variables is closely similar to that described in Table 5.12, the main differences being in separating some predictors (for example, plage compactness from filament in parameter #10). Neidig et al. omitted parameter #23, daily 10-cm flux and added a type of radio burst. Vecchia et al. added a recoded sunspot descriptor, and rescaled a number of variables.

The analyses of Hirman et al. (1980) and of Neidig et al. (1981) included six combination variables, described in Table 5.13. As discussed in Chapter 4, these combination variables were among those selected by the analytical procedure as containing information useful for forecasting flares. Vecchia et al. (1980) eschewed the use of arbitrary combinations. The studies differ, also, in the data used

TABLE 5.13

COMBINATION PARAMETERS USED BY HIRMAN ET AL. (1980) AND BY NEIDIG ET AL. (1981)

Parameter No.	Parameter Formula
New #1	product of 3 sunspot descriptors
New #2	New #1 x #2
New #3	New #1 x #16
New #4	#3 x #4 (#5 + #11)
New #5	#2 x #5 x #12
New #6	#5 x (#11+#12+#13)

*Numbers in "Parameter formula" correspond to parameter numbers in Table 5.12.

for training and for verification. The various methods tend to select the same parameters--with a choice of 22 similar predictors, among the first 14 selected in each of the three studies, 11, or approximately 80% were in common. The order of selection was quite different in the three studies, however. According to the results of these similar studies of a limited part of the sunspot cycle, predictions based on about half the number of predictors would be about as successful as those based on the full complement of data. One needs to keep in mind, however, that different predictors are useful for different levels of activity, and for predicting large and small flares, according to Jakimiec and Wasiucionek (1980).

Summary

What can we make of these extensive and varied lists of flare predictors? First, we attempt to fit each criterion or predictor into one of four classes (generating yet another list). The usefulness of this exercise is to see the relations and redundancy among predictors and to simplify what may at first appear to be chaos. We shall see that there probably is, after all, some physical basis for flare forecasts.

Table 5.14 contains most of the predictors in Tables 5.7 through 5.12. The collection of predictors can be seen

TABLE 5.14

SUMMARY OF FLARE PREDICTORS

I. Persistence.
- Occurrence of flares, sprays, proton flares, X-ray bursts, radio bursts;
- H-alpha bright points;
- Active longitudes.

II. Magnetic Field

A 1. Strength: condition
- Field strength (usually line-of-sight component);
- Transverse and total fields;
- Sunspot area; sunspot penumbra.

A 2. Strength: evolution.
- Growth, decay of magnetic field;
- Growth of region, spot, penumbra.

B 1. Stress: condition.
- Spatial gradient of field, magnetic complexity, delta configuration or satellite spot, neutral-line kinks.
- Magnetic shear, transverse field parallel to neutral line, filament-aligned fibrils;
- East-west neutral line, reversed polarity.

B 2. Stress: evolution.
- Growth, decay of magnetic-flux cells, plage, spots;
- Spot birth, emerging-flux region;
- Velocity shear, sunspot proper motions (shear or rotation);
- Region interaction: colliding regions, birth of new region within an existing region.

III. Enhancement of density and temperature in active-region.
- Region brightness and area: H-alpha plage, CaII K plage, microwave flux, soft X-ray flux, coronal green-line intensity.
- Temperature: microwave spectrum, X-ray spectrum, coronal yellow line.

IV. Mass motions: Filament activation, surges, moustaches, coronal mass ejection.

as indicators of (1) enhanced rate of occurrence of flares and flarelike events; (2) strong, stressed magnetic field; (3) growth and active-region dynamics that lead to magnetic stress; or (4) enhanced density and temperature in magnetically contained plasma. In short, practical predictors of flares describe an active region's performance or its potential performance as a flare producer in terms that are physically reasonable, although not always direct and precise.

The predictors that consistently appear as most informative in objective and subjective predictions are persistence of flare activity and measures of magnetic-field strength, gradient and complexity. By providing easy intercomparison, this grouping of predictors could help reduce redundancy.

The next section describes the product of these efforts of gathering, sharing, and interpreting information.

5.4 MAKING THE FORECAST

The five forecasters at SESC "rotate with the Sun." Working 5-day shifts, they are locked into a 25-day pattern close enough to the Sun's 27-day rotation period that each forecaster becomes familiar with the recent history of activity within certain ranges of heliolongitudes. The forecaster studies each active region on the Sun, analyzing as many as 40 parameters. Sketches of sunspots and of H-alpha filaments are combined with quantitative data on magnetic-field strengths and polarities to show gradients and neutral lines and illustrate evolutionary trends.

Four separate steps or points of view are considered in developing the SESC forecast for each active region:

1. The "region-type forecast" is based on region statistics; for example, sunspot groups of the most active class have historically produced 3 large (M or X) bursts per day.
2. The "trend forecast" considers how the region is changing, and extrapolates its development to each of the next 3 days.
3. "Persistence" modifies the forecast in accordance with this particular region's record as a flare producer.
4. An "analogy forecast" is based on the outcome of a similar situation that occurred in the past.

This description is based on a presentation by Hirman (1985), but nomenclature has been modified, reserving the term "climatology" to mean a long-term average, appropriate to a given phase in the solar cycle. Unbiased forecasts would begin at that level.

Hirman described forecasting as one of the most difficult and scientifically demanding tasks attempted on a routine basis. The forecaster, he said, is "faced with the challenge of bringing to bear, on a set of incomplete and error-prone initial data, whatever may be relevant from a vast and complex science, within strict deadlines, and of retaining the confidence to do so day after day with only rarely the chance to go back and try to find out why he might have been right or wrong."

At SESC, forecasts are subjective, and individuality is cherished. There is a tradition that forecasts may be evaluated, but not forecasters. The need for numerical guidance is often mentioned.

5.5 AN OPERATIONAL SYSTEM OF NUMERICAL FORECASTING

A computer-based system for data gathering and display, numerical prediction, and forecast evaluation is operating in Warsaw. The Helio-Geophysical Prediction Service at the Space Research Center of the Polish Academy of Sciences is an Associated Warning Center of IUWDS. The following description is based on an informal report made available at the Meudon Workshop on Solar Activity Prediction (Klos and Stanislawska 1984).

The minimum equipment necessary to operate the system, called HELGEO, consists of a minicomputer with 32k memory, magnetic tape or disks for storage, and telephone modem, plus teletype and access to the IUWDS data network. The flow from data reception to forecast and verification includes the following steps:

1. Coded data (URSIgrams) received by teletype are entered on punched tape or magnetic tape. Local data can be entered in a similar way.
2. The data are read into the computer, checked, and corrected.
3. The corrected data are decoded and tabulated by the computer, and can then be displayed or printed. A single sequence of many subprograms then orders and averages the decoded data, prepares a report of

solar and geophysical activity, and a forecast of solar and geophysical conditions and events.

4. The message is transmitted to users by teletype or telephone.

Another sequence of computer programs examines the correctness of each prediction.

The system's modular construction facilitates improvement through replacement or addition of modules. Future plans include a block of modules to correct errors in the coded data received from the IUWDS World Warning Center (SESC) and from the Regional Warning Centers.

The decoded, averaged, and thematically ordered data can be displayed in tabular form. A subroutine HELIOMAP presents a graphical display of photospheric features and another display of flares.

Users can receive the product by teletype, or by telephone at their own computer terminals. Like the daily report from SESC, HELGEO's product consists of two parts: reports and forecasts. The first part includes reports of optical, X-ray, and radio flares; other solar activity; geomagnetic and ionospheric conditions and disturbances; and radio-wave propagation conditions for various paths and times. The second part contains predictions of X-ray flares; probability of optical flares; and probability of SWF occurrence, severity, and duration. It also includes predictions of geomagnetic and ionospheric storms and indices, and of radio-wave propagation conditions over several paths.

Each of these predictions is compared to the outcome, and statistics on these differences are gathered and displayed by the sequence of computer programs designed for forecast evaluation. Thus the system provides for its own assessment, as well as for continual modification and improvement.

5.6 FORECASTER NEEDS

Other forecaster needs, in addition to numerical guidance, were expressed at meetings of the Forecasters Working Group at Meudon. These include improvement of research input, forecast evaluation, and more sharing of ideas and techniques among centers.

Pleas for improved interaction between researchers and forecasters noted that it takes 10 years for research to find application in operations. One forecaster said that "what we get [from research] bears no relation to what we need." The group concluded that "forecasters and researchers should establish more day-to-day contact."

There was a demand for some uniform way to verify and evaluate forecasts, for routine comparison of forecast methods and results, even for "an international standard against which we can judge our models." [Clearly, uniform verification and evaluation will await standardization of the quantity forecast.] Forecasters expressed the need to improve communication among forecast centers, and to share and examine the rationale behind the forecast.

Is training one of the needs of a solar forecaster? The answer depends on the respondent. Experience, rather than formal training, is what is needed according to Hirman, who points out that senior forecasters have a background of solar observing or weather forecasting, and have worked throughout a complete 11-year solar cycle. USAF forecasters have a 4-year tour; NOAA-Corps officers stay only 2 years. Journeyman forecasters at SESC are relative newcomers. Others help out at the forecast desk during times of high activity or special missions. Hirman says the senior forecasters can recognize and correctly categorize most developing solar situations and can answer almost any question that a user may have. A journeyman forecaster can handle operational duties, but when something unusual comes up must rely on someone with more experience.

Thus, at SESC, training is nearly synonomous with experience. Forecasters at the European centers tend to have more formal scientific training. Does this make a difference in the quality of the forecast? One research scientist at SEL maintains vigorously that it does. At present, however, there are many other differences among centers and there is no comparison of forecasts, so there is no way to determine this.

Hirman said that "forecasting is not an exercise in physics and applied mathematics, it's an exercise in recognizing, recalling, categorizing, and decision making," but he concludes, nevertheless, that "the forecaster needs all the help he can get."

From what has been said so far, it seems clear that the primary need is for good data--quantitative, accurate, timely, and complete. If objective numerical analysis derives little useful information from incomplete or qualitative data sets, it is hard to see how a forecaster can get more out of the same data.

The program SELDADS II (Space Environment Laboratory Data Analysis and Display System; Table 5.5) is designed to meet the need for timely data. The set of image-handling algorithms planned for the system is called SELSIS--Space Environment Laboratory Solar Image System. It will bring to the forecaster's console near real-time H-alpha images from the SOON sites and longitudinal magnetic-field images from the National Solar Observatory at Kitt Peak.

Another clear need is for realistic evaluation of forecasts (the subject of Chapter 6). Without this, assessment of forecast success is only a guess. One benefit of adequate evaluation would be a practical view of what success is possible, given the data and understanding that exist. A search for good evaluation would include a search for the format that would give best expression to the skills solar forecasters have (identifying active regions likely to flare, assessing the probability of a flare, and estimating flare size), and would de-emphasize the skill they lack, which is to predict the time of the flare within periods shorter than a few days.

Forecasters plead for help in dealing with the data flood. One way would be to identify and use only a few of the most relevant parameters; another would be to program the computer to analyze, select, and combine all the incoming information. This would require entering, verifying, storing, and retrieving large sets of data. Availability of this data base to the forecaster may be an important part of the aid that numerical guidance could provide.

Some things forecasters don't need are "promising" unconfirmed research results, overwhelming concern for the consequences of forecast misses and false alarms, dichotomies between objective and subjective techniques or between physics and practice, and negative skill scores.

5.7 REFERENCES

Avdyushin, S.I., N.K. Pereyaslova, F.L. Dlikman, and Yu.M. Kulagin, 1979. Forecasting of solar energetic radiation at the forecast center of the Institute of Applied Geophysics, pp. 89-103 in Solar-Terrestrial Predictions Proceedings, Vol. I: Prediction Group Reports, R.F. Donnelly (ed.). U.S. Dept. of Commerce, NOAA, ERL, Boulder, Colo.

Bartkowiak, A., and M. Jakimiec, 1986. Relationships among characteristics describing solar active regions. Proc. of Meudon Solar-Terrestrial Predictions Workshop (1984), G. Heckman, M.Shea, and P. Simon (eds.), Boulder, Colo.

Burov, V.A., J.W. Hirman, and W.E. Flowers, 1980. Forecasting flare activity by pattern recognition technique, pp. C-235-C-241 in Solar-Terrestrial Predictions Proceedings, Vol. III: Solar Activity Predictions, R.F. Donnelly (ed.). U.S. Dept of Commerce, NOAA, ERL, Boulder, Colo.

Chen, X.-Z., and A.-D. Zhao, 1979. Short-term solar activity forecasting, pp. 176-181 in Solar-Terrestrial Predictions Proceedings, Vol. I: Prediction Group Reports, R.F. Donnelly (ed.). U.S. Dept. of Commerce, NOAA, ERL, Boulder, Colo.

Ding, Y.-J., and Zhang, B.-R., 1979. Spiral sunspots and solar activity forecasts, pp. 140-153 in Solar-Terrestrial Predictions Proceedings, Vol. I; Prediction Group Reports, R.F. Donnelly (ed.). U.S. Dept. of Commerce, NOAA, ERL, Boulder, Colo.

Dodson, H., and E.R. Hedeman, 1972. Large-scale organization of solar activity in time and space, pp. 19-31 in Solar Activity Observations and Predictions, P. McIntosh and M. Dryer (eds.). M.I.T. Press, Cambridge, Mass. (Progress in Astronautics and Aeronautics, vol. 30).

Enome, S., 1979. A review of short-term flare forecasting activities at Toyokawa, pp. 205-211 in Solar-Terrestrial Predictions Proceedings, Vol. I: Prediction Group Reports, R.F. Donnelly (ed.). U.S. Dept. of Commerce, NOAA, ERL, Boulder, Colo.

Heckman, G., 1979. Predictions of the Space Environment Services Center, pp. 322-349 in Solar-Terrestrial Predictions Proceedings, Vol. I: Prediction Group Reports, R.F. Donnelly (ed.). U.S. Dept. of Commerce, NOAA, ERL, Boulder, Colo.

Hirman, J., 1985. Presentation at Workshop on Solar-Flare Prediction, Stanford, Calif.

Hirman, J., D.F. Neidig, P.H. Seagraves, W.E. Flowers, and P.H. Wiborg, 1980. The application of multivariate discriminant analysis to solar flare forecasting, pp. C-64-C-75 in Solar-Terrestrial Predictions Proceedings, Vol. III: Solar Activity Predictions, R.F. Donnelly (ed.). U.S. Dept. of Commerce, NOAA, ERL, Boulder, Colo.

Holloman Solar Observatory, 1982. Solar Observing Optical Network. Det.4, 4th Weather Wing, Holloman AFB, N.M.

International URSIgram and World Day Service, 1973. IUWDS Synoptic Codes for Solar and Geophysical Data, 3rd rev. ed. NOAA, SEL, Boulder, Colo.

Klos, A., and I. Stanislawska, 1984. Informing-predicting

system/IPS/HELGEO. Informal rept. from Helio-Geophysical Prediction Svc., Polish Acad. of Sciences, Space Res. Ctr., Ordona 21, 01-237, Warszawa, Poland.

Jakimiec, M., and J. Wasiucionek, 1980. Short-term flare predictions and their stationarity during the 11-year cycle, pp. C-54-C63 in Solar-Terrestrial Predictions Proceedings, Vol. III: Solar Activity Predictions, R.F. Donnelly (ed.). U.S. Dept. of Commerce, NOAA, ERL, Boulder, Colo.

Krivsky, L., 1981. Organization and methodology of weekly predictions of solar activity in Czechoslovakia. Phys. Solariterr., Potsdam, no. 16, 95-105.

Neidig, D., P. Wiborg, P. Seagraves, J. Hirman, and W. Flowers, 1981. An objective method for forecasting solar flares. AFGL-TR-8-0026, Air Force Geophysics Lab., Hanscom AFB, Mass.

Peking Observatory, Solar Activity Prediction Group, 1979. Solar activity predictions at Peking Observatory, pp. 154-162 in Solar-Terrestrial Predictions Proceedings, Vol. I: Prediction Group Reports, R.F. Donnelly (ed.). U.S. Dept. of Commerce, NOAA, ERL, Boulder, Colo.

Severny, A.B., N.N. Stepanyan, and N.V. Steshenko, 1979. Soviet short-term forecasts of active region evolutions and flare activity, pp. 72-88 in Solar-Terrestrial Predictions Proceedings, Vol. I: Prediction Group Reports, R.F. Donnelly (ed.). U.S. Dept. of Commerce, NOAA, ERL, Boulder, Colo.

Shapley, A., 1973. Introduction, pp. 7-9 in IUWDS Synoptic Codes for Solar and Geophysical Data, 3rd rev.ed. NOAA, SEL, Boulder, Colo.

Simon, P., 1973. Some comments on the forecasting of solar activity, pp. 11-19 in IUWDS Synoptic Codes for Solar and Geophysical Data, 3rd rev. ed. NOAA, SEL, Boulder, Colo.

Simon, P, 1979. The forecasting center of Meudon, France, pp. 1-11 in Solar-Terrestrial Predictions Proceedings, Vol. I: Prediction Group Reports, R.F. Donnelly (ed.). U.S. Dept. of Commerce, NOAA, ERL, Boulder, Colo.

Simon, P., J.B. Smith, Jr., Y. Ding, W. Flowers, Q. Guo, K. Harvey, R. Hedeman, S. Martin, S. McKenna-Lawlor, V. Lin, D. Neidig, V. Obridko, H. Dodson-Prince, D. Rust, D. Speich, A. Starr, and N. Stepanyan, 1979. Short-term solar activity predictions, pp. 287-321 in Solar-Terrestrial Predictions Proceedings, Vol. II: Working Group Reports and Reviews, R.F. Donnelly (ed.). U.S. Dept. of Commerce, NOAA, ERL, Boulder, Colo.

Solar Activity Prediction Group, Department of Solar Physics, Peking University, 1979. Solar activity predictions at Peking Observatory, pp. 154-162 in Solar-Terrestrial Predictions Proceedings, Vol. I: Prediction Group Reports, R.F. Donnelly (ed.). U.S. Dept. of Commerce, NOAA, ERL, Boulder, Colo.

Stasiewicz, K., M. Maksymienko, and M. Jakimiec, 1979. KAZIA--A computer system for solar-terrestrial data processing and objective prediction, pp. 61-66 in Solar-Terrestrial Predictions Proceedings, Vol. I: Prediction Group Reports, R.F. Donnelly (ed.). U.S. Dept. of Commerce, NOAA, ERL, Boulder, Colo.

Svestka, Z., 1972. Solar Flares. Dordrecht: D. Reidel.

Vecchia, D.F., P.V. Tryon, G.A. Caldwell, and R.H. Jones, 1980. Statistical methods for solar flare probability forecasting. AFGL, Hanscom AFB, Mass.

Zirin, H., 1983. The 1981 July 26-27 flares: Magnetic developments leading to and following flares. Astrophys. J. **274**, 900-909.

6. *Forecast Evaluation*

6.1 INTRODUCTION

Forecasters at the 1980 Meudon workshop expressed the need to compare and evaluate forecasts. Without realistic evaluation there is no way to measure progress or to compare methods. Caution is indicated, however: a score that fails to measure the real value of the forecast can do more harm than good. Realistic evaluation, if attainable, would identify successful individuals, who could then be encouraged to share their insights. It might show whether one forecasting center has something to learn from another; it might make the job more fun. Forecasters need to know how they are doing, and users would benefit from information about the accuracy and limitations of the forecasts they must depend on in making operational decisions. However, defining an adequate method of forecast evaluation may be a tough job.

A basic problem in comparing one forecast center to another is that different centers often forecast different events or conditions. When they do forecast the same type of event, they may describe it differently. Another problem lies in the fact that present knowledge enables forecasters to identify an active region that is likely to flare, but not to pinpoint the time of eruption: the timing uncertainty of 2 or 3 days is greater than the forecast interval. There is a third intrinsic problem. Rare events are especially difficult to predict, but rare events are what solar forecasters must deal with. In 1967, for example, there were 5 days with a flare of importance 3, whereas only 77 days lacked a flare of importance 1 or greater. Thus days with a big flare and days with no flare were both unusual events. A final problem to discuss is identification of a suitable standard to measure forecasting skill.

6.2 EXAMPLES OF FORECAST VERIFICATION

The timing problem was vividly described by the Working Group on Short-term Solar Activity Predictions at the 1979 workshop. Meudon once forecast "active" conditions on four successive days; when a large flare finally occurred on the fourth day, the forecast was considered a true success because no one knows how to predict the exact time of a flare. "One can compute the actual probability but it would be very difficult to understand why the forecast was a failure." (Simon et al. 1979, p. 315). Other examples of the right forecast at the wrong time appear in Figure 6.1, a timeline for the last quarter of 1967 (Gray and Slutz 1967). December 1 is one example when a flare occurred the day after a flare was forecast. At the end of October flares started erupting after a barren period with elevated forecast probabilities. A score based on daily comparison of forecast to outcome makes a timing error of one day very costly, adding once to the "miss" count and once to the "false-alarm" count.

The problem of forecasting rare events is illustrated by Table 6.1, the verification matrix for daily yes/no forecasts of flares of importance 2 or greater in 1967. Entries are the number of days that fit each situation. For example, flares were forecast for 71 days, and on 18 of these, a flare occurred. In parentheses are the numbers expected if the forecasts were unrelated to flare occurrence: 9 = (71x45)/365. Given the totals, 71 and 45, the probability of getting 267 + 18 = 285 forecasts correct by chance is less than one in a thousand, according to a standard (chi-square) test. This indicates that the forecasts express a good deal of useful information. Forecasters identified days when flare rate was enhanced, although the rate reached only 25%.

In this example, 285, or 78%, of the forecasts are correct. If a forecast of "no flare" had been issued each day, 88% (320/365) would have been correct. As a score for forecasts of rare events, the fraction of forecasts correct is uninformative. Even worse, it is "improper" because it discourages forecasters from expressing their best judgment.

At the 1979 Boulder workshop, "fraction of forecasts correct" was the typical expression of forecast reliability. This was true even of Simon's (1979, p. 10) report for Meudon forecasts, although earlier in the presentation (p. 2), Simon pointed out the absurdity of this criterion. Heckman (1979a) was another participant who emphasized the need for a "proper" score.

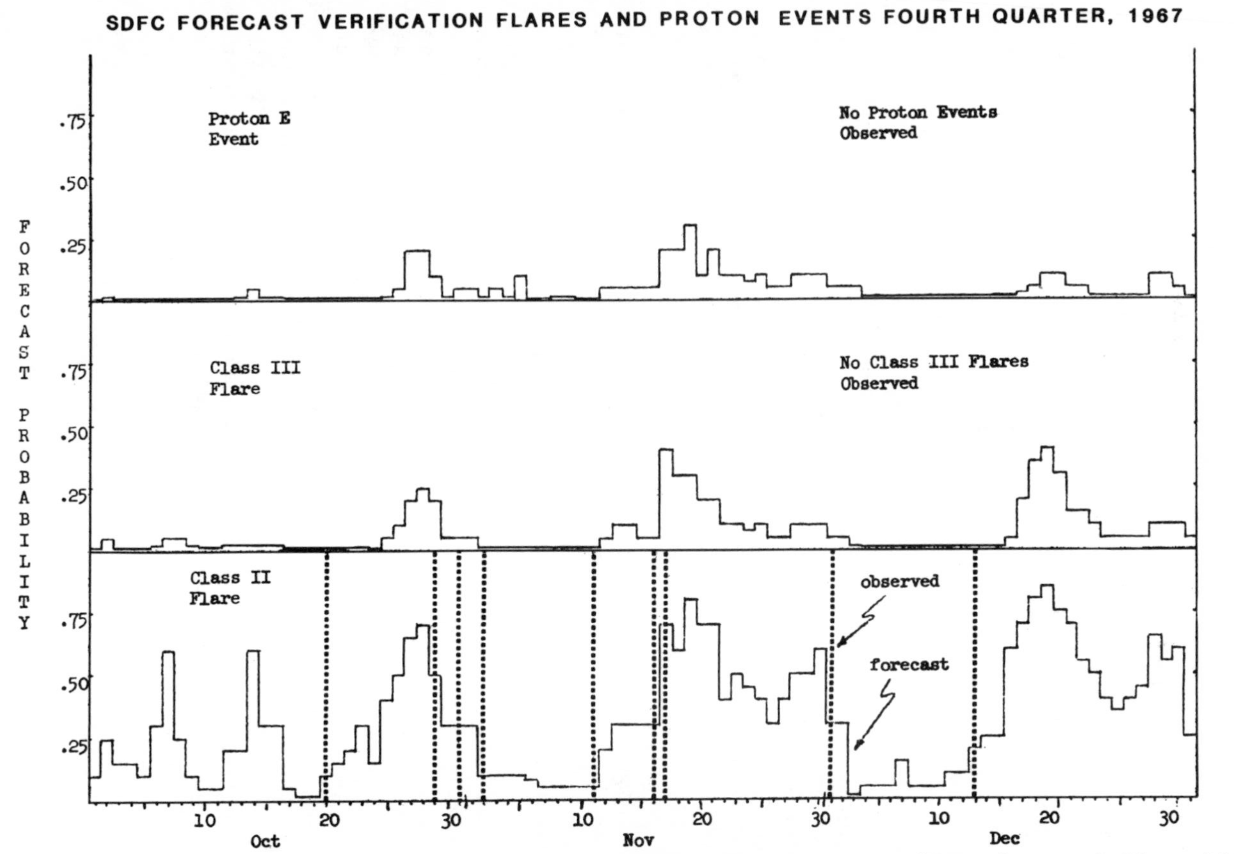

Figure 6.1--Timeline of SESC forecasts of X-ray bursts and of outcome. (From Gray and Slutz 1967).

TABLE 6.1

VERIFICATION MATRIX
FOR FORECASTS OF FLARES OF IMPORTANCE >2

OUTCOME:	FORECAST: "NO FLARE" (0)	"FLARE" (1)	TOTAL
NO FLARE	correct "quiet" forecast 267 (258 expected)	false alarm, overforecast 53 (62 expected)	320
FLARE	miss, underforecast 27 (36 expected)	correct "flare" forecast 18 (9 expected)	45
TOTAL	294	71	365

Figure 6.2 is another example that shows the amount of information provided by solar-burst forecasts and the difficulty of predicting rare events. A set of forecasts and outcomes for X-ray bursts is shown in the form of a bar graph. Daily active-region forecasts and outcomes are taken from Table 3 of Neidig et al. (1981). Grouping is determined by the forecast of "no burst" (left), "C" (middle), or "M or X" (right). For each group, the outcome is shown as the number of occasions when bursts actually occurred, and as the percent of all forecasts in that group. The graph illustrates (1) the relatively small number of forecasts of bursts of class M or X and the even smaller number of actual occurrences; (2) the considerable amount of information provided by the forecasts, demonstrated by the relatively large proportion of M,X occurrences in the "M,X forecast" group; and (3) the fact that, nevertheless, because of the overwhelmingly large number of days with "no M or X" forecast, among M,X that occurred, fewer (63) were forecast than not (84), and among forecasts of "M or X", fewer (63) occurred than not (159). The high rate of "misses" and of

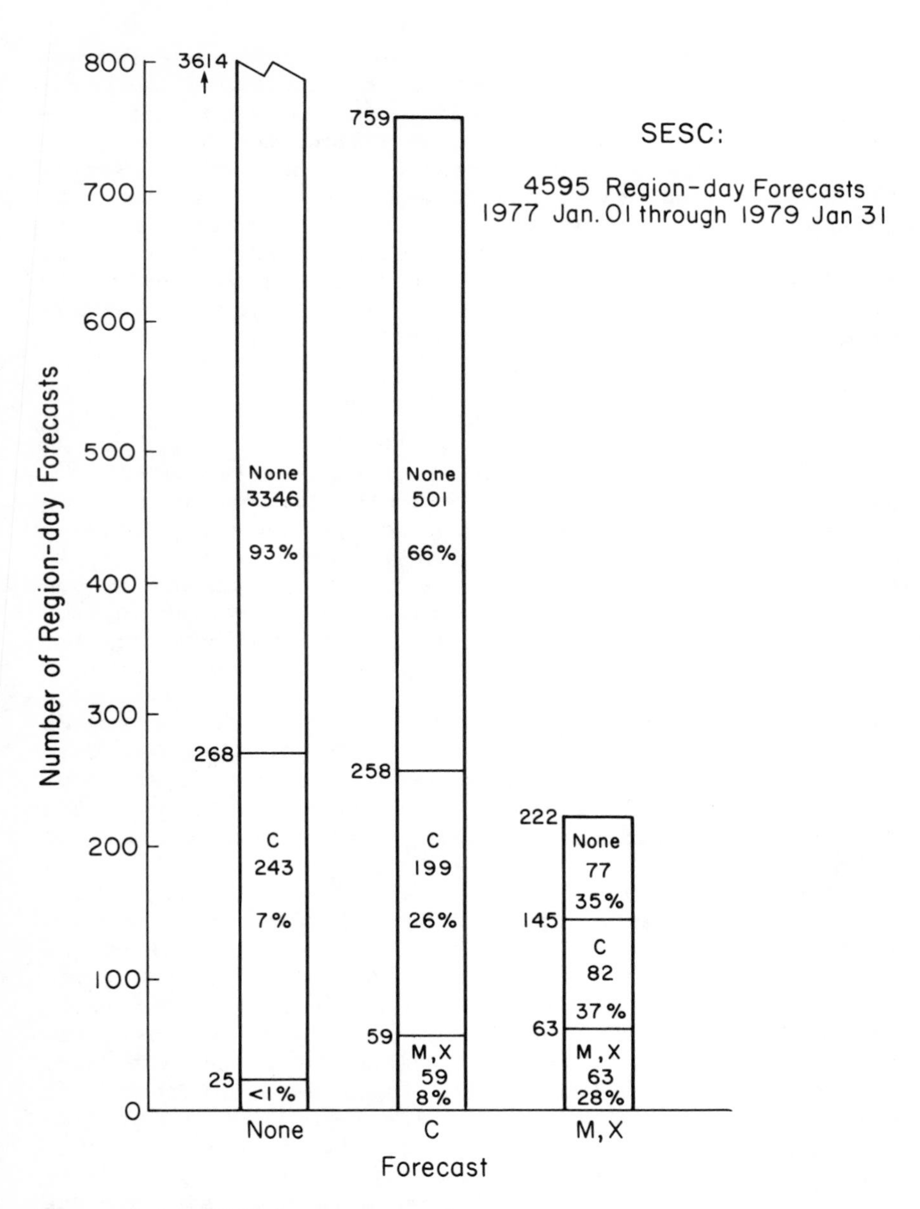

Figure 6.2--SESC forecasts and outcome.

"false alarms" is expected in forecasting rare events. Again, there is the brutal fact that only 79% of the forecasts are correct, whereas a mindless but consistent forecast of "no M or X burst" would be 97% correct.

Can we hope to find a better way to summarize information on the accuracy of the forecast and the skill of the forecaster? It is easy to see that "fraction of forecasts correct" is a poor score, but not easy to see what would be a good score. In the next section the scores used by weather forecasters will be examined and applied to a solar example. This will raise more questions.

6.3 SCORING PROBABILITY FORECASTS

The final measure of forecast success is, of course, based on utility. How useful is the forecast in guiding decisions that must be made in carrying out a practical operation? The user can assign an economic gain or loss to the outcome in each cell of the verification matrix and combine these weights with the number of forecasts in each cell to assess the overall utility of the forecasts to that particular operation. But the great variety of uses of flare forecasts makes it impractical for the forecaster to take into account the probable gains and losses for each type of user, so the two parts of the problem are kept separate. The forecaster's concern is to provide an accurate estimate of the probability of flare occurrence. The economic, scientific, or other practical weights can be added by the user in a separate step.

Solar forecasters at SESC, like meteorologists, make a probability forecast. This allows a forecaster to express an estimate of what will happen and also an estimate of the uncertainty of the first estimate. Most people are comfortable with the idea of an average value with attached probable error, and most can interpret weather forecasts expressed as probability (with occasional whimsical reference to the effort involved in clearing away a 10% chance of snow). Solar-burst forecasts have a similar format, expressing for a given day the probability that a burst with peak flux above a given threshold will occur.

Solar forecasters at SESC, like meteorologists, use the Brier F score: the average squared difference between forecast and outcome. It has the dimensions of dispersion or squared variance, but the forecast is compared to the outcome in each case, rather than to the mean. Being the sum of squares, the Brier F score is never less than 0; forecasts close to reality yield low F scores. Two advantages are cited for the F score: (1) It is a "proper score" that rewards expression of the best estimate of occurrence probability (Murphy and Epstein 1967). (2) Although the measure of utility depends on application, the F score in any case provides correct ranking of utility of two-category forecasts (Hunt 1963; Murphy 1970).

The mean occurrence frequency of the forecast event is called "climatology" in solar forecasting, as in weather forecasting. The "climatology score," C, corresponds to a uniform forecast of this mean value, climatology, every day. The F score is compared to C in the "skill score," S = (C − F)/C. Advantages of S are that it has a defined range of possible values (−1 to 1), and that it follows the more familiar convention that big is good.

For solar forecasting, however, defining climatology is far from trivial. The eleven-year activity cycle imposes on flare occurrence frequency a variation of large amplitude and only partial predictability. Advance knowledge of solar climatology amounts to a long-term forecast. What value of flare occurrence frequency should be used to estimate C in the score? Use of actual climatology for the scoring period gives the C forecast an advantage of hindsight that amounts to a severe handicap for the actual forecaster. On the other hand, use of the forecast climatology removes from the score any information about skill in long-range forecasting. The average probability for that phase of the activity cycle would be most appropriate but requires historical data on occurrence frequency of the forecast quantitity, in this case, X-ray bursts.

In the following examples the overall mean outcome (Mean AO = 0.1714 in Table 6.2) is used as climatology. Table 6.2, derived from Table B0 of Vecchia et al. (1980), presents 4487 region-day forecasts of the probability of occurrence of an X-ray burst of magnitude C or greater. These forecasts were made at SESC in 1977 and 1978. Similar forecasts are grouped together: probability <0.1, 0.1 to 0.2, . . . , 0.9 to 1.0. The number of region-days that actually produced bursts is noted for each forecast group. Individual outcomes are categorical: a region-day with at least one burst is counted as 1, a region-day with no burst as 0. The mean outcome for each group is just the proportion of region-days with a burst.

Perfect forecasts would fall into two groups, one with mean outcome 1, labeled 1, the other with mean outcome 0, labeled 0. In fact, of course, forecast and outcome are neither identical nor unrelated. This is seen in Figure 6.3

TABLE 6.2

EXAMPLE OF F SCORE AND SKILL SCORE APPLIED TO SESC PROBABILITY FORECASTS
(Entry is number of region days in each category)

FORECAST:	0.05	0.15	0.25	0.35	0.45	0.55	0.65	0.75	0.85	0.95	Total
OUTCOME:											
No burst, M0	2171	426	270	194	202	101	80	115	132	27	3718
Burst, M1	126	48	58	42	51	52	41	74	191	86	769
Total, M	2297	474	328	236	253	153	121	189	323	113	4487
Mean AO:	0.055	0.101	0.177	0.178	0.202	0.340	0.339	0.392	0.591	0.761	0.1714 = r = 769/4487 = "climatology forecast"
AO(1-AO):	0.0519	0.0910	0.1456	0.1463	0.1609	0.2244	0.2240	0.2382	0.2417	0.1818	0.1093 (sharpness) 0.0209 (labeling)
Score F:	0.0519	0.0934	0.1509	0.1759	0.2227	0.2685	0.3208	0.3667	0.3086	0.2175	0.1302 = F
Score C:	0.0654	0.0959	0.1456	0.1463	0.1619	0.2527	0.2521	0.2867	0.4180	0.5296	0.1420 = C
C - F:	0.0136	0.0025	-0.0053	-0.0296	-0.0608	-0.0158	0.0688	0.0800	0.1094	0.3120	0.0118 = C - F
Skill score:	0.208	0.026	-0.036	-0.202	-0.376	-0.063	-0.273	-0.279	0.262	0.059	0.083 =(C-F)/C

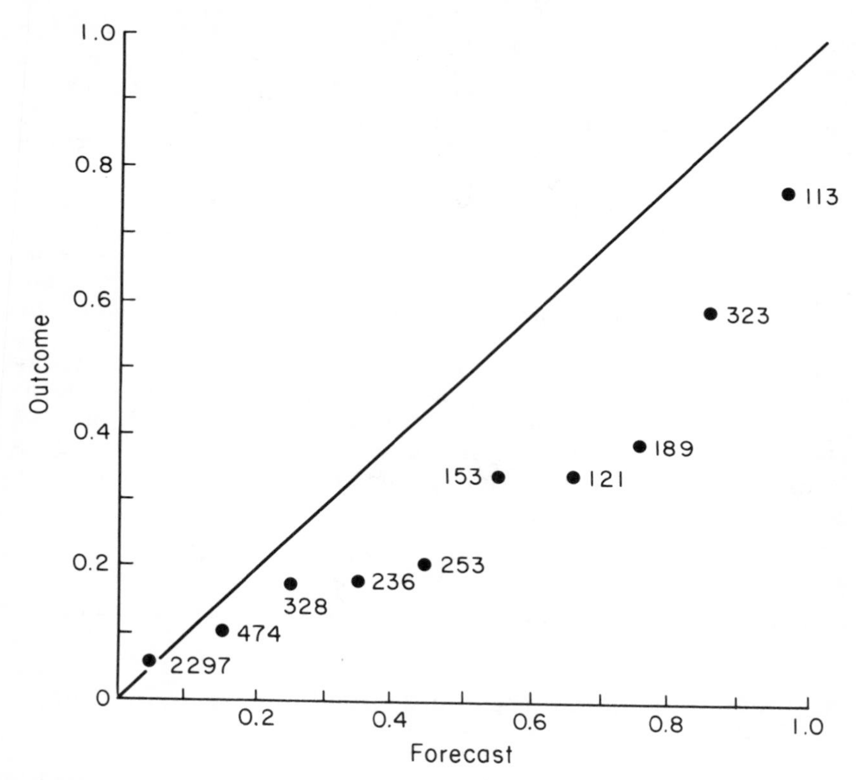

Figure 6.3--Mean outcome plotted against forecast for region-day probability forecasts grouped by forecast probability; data are from Table 6.2.

where the (f, AO) pair is plotted for each of the ten forecast groups. Occurrence frequency does increase as forecast probability increases, but the forecast probability tends to be systematically higher than the outcome. It suggests that it would be useful to measure separately the sorting skill shown by the definite trend of the points in Figure 6.3, and the bias shown in labeling the forecast groups with probabilities that are systematically higher than the actual occurrence frequencies, the outcome.

Sanders (1963, 1966) made just this distinction, expressing scores as the sum of two components. The first component measures the ability to recognize high probability of occurrence and to sort situations into groups with outcomes that differ from one group to another, and that are homogeneous within each group. This sorting skill has also been called "sharpness": it measures how clearly forecasts differentiate outcomes. Sanders called the second skill "labeling skill." It is the ability to label each forecast group with a realistic probability value. Bias in the forecasts shows lack of labeling skill.

Tables 6.3 and 6.4 are simplified examples constructed in order to show the effect of sorting and labeling skills on the F score and on the skill score. The five groups of Table 6.2 with low probability forecast are merged into one group labeled 0 in Table 6.3, and the rest into a second group labeled 1. Then F = (708 + 325)/4487 = 0.230. In this case, with groups labeled 0 and 1, F is just the fraction of incorrect forecasts, or the fraction of the population corresponding to the sum of the 325 misses and the 708 false alarms. In these examples, the skill score S is calculated with the climatology score C = 0.1420 from Table 6.2.

TABLE 6.3

CATEGORIZED FORECAST LABELED 0, 1; F = 0.230, S = -0.621

	Forecast probability: <0.5 (0)	>0.5 (1)	Total
Outcome: NO FLARE (0)	3010=67%	708=16%	3718= 83%
FLARE (1)	325= 7%	444=10%	769= 17%
Total	3335=74%	1152=26%	4487=100%

Next, we can improve labeling skill by giving each of the two groups a more precise label: the average probability forecast for that group. Entries in the contingency table do not change, but the forecast probabilities 0 and 1 are replaced by the average values, 0.139 and 0.596. Now

$$4487\ F = 3010(0.139) + 325(1 - 0.139) + 708(0.596) + 444(1 - 0.596),$$

and F = 0.139.

TABLE 6.4

CATEGORIZED FORECAST, REALISTICALLY GROUPED
PRECISELY LABELED; F = 0.129, S = 0.094

Outcome:	Forecast: <0.8 (0.1898)	>0.8 (0.8759)	Total
NO FLARE (0)	3559	159	3718
FLARE (1)	492	277	769
Total	4051	436	4487

The new value of F is smaller than C, and the skill score S improves rather dramatically, increasing from a large negative value to a small positive value (S = 0.022).

In Table 6.4, the division of forecast groups is chosen to emphasize sorting skill. The data in Table 6.2, plotted in Figure 6.3, show that actual occurrence frequency or outcome is distinctly higher for the forecast groups labeled 0.8 and 0.9 than for the other forecast groups. When the forecasts are divided into two groups separated at forecast probability 0.8, scores again improve.

Finally, if we use hindsight to assign to each forecast group realistic labels, 0.1215 and 0.6353, that describe the actual outcome, we find F = 0.119 and S = 0.163.

In reviewing these results, recall that $F \geq 0$, and a perfect score would be F = 0, whereas $-1 \leq S \leq 1$, and a perfect score would be S = 1. Then for:

2 classes, divided at 0.5, labeled 0,1:	F= 0.230, S =-0.621;
the same classes, precisely labeled:	F= 0.139, S = 0.022;
divided at 0.8, precisely labeled:	F= 0.129, S = 0.094;
divided at 0.8, realistically labeled:	F= 0.119, S = 0.163.

Score improvement from line 2 to line 3 is due to improved skill in sorting situations into groups with distinctly different outcome. The other examples illustrate the sensitivity of the F score and the skill score to unbiased and precise labeling of the forecasts. The last line shows how scores would improve if bias could be removed and the forecast groups labeled with probabilities that agree with actual occurrence frequencies.

The remaining part of the F score, 0.119, must be attributed to lack of sorting skill. For the fully sorted data in Table 6.2, lack of sorting skill or sharpness contributes 0.109 to the F score--forecasts in ten groups are sharper than in two. Lack of sharpness accounts for 84% (0.109/0.130) of the total F score. The other 16% of the total F score is "bias penalty" due to lack of skill in assigning a realistic probability to each forecast group. If each group were perfectly labeled so that forecast probability equaled outcome occurrence frequency, this part of F would disappear. The 16% contribution to F is large compared to the difference (C - F), and it lowers the skill score from S = 0.23 = (0.142-0.109)/0.142 without bias or labeling error to S = 0.08 with bias. Thus, the skill score exacts a high penalty for systematic bias. We speculate that the persistent bias toward forecasting unrealistically high probabilities arises from the forecaster's perception that a miss is more costly than a false alarm to the user.

Inhomogeneity of outcome within groups and lack of distinction between groups is measured by the mean squared deviation of individual outcomes from the mean outcome for the group. Because outcomes, AO, are 0 or 1, this is simply the mean value of the product AO(1 - AO) for a group. When outcome is homogeneous within a group, AO is near 0 or 1, and the product AO(1 - AO) is small. Thus, the middle forecast groups, with a mixture of outcomes, contribute most to the F score. Sanders (1966) made the point (in terms of meteorological forecasts) that a forecaster can better the score by minimizing the number of forecasts in the uncertain midrange of probabilities, but not beyond the point where the labeling or bias part of the score penalizes overconfidence. The score rewards the ability to recognize and label situations with the most certain outcome.

The chi-square test, in contrast to low skill scores, indicates a close relation between forecasts and outcomes in this example: the odds of getting as close a relation by chance are much less than 1 in 10,000. Then why is the skill score so low? Going back to the 1967 example in Table 6.1, we find that the F score for the forecasts (with chi-square probability <0.001) is worse than the climatology C score, yielding a negative skill score. The forecast probability is about half again greater than the outcome--the forecasts are biased toward high values. With a climatological forecast as the standard, the bias leads to a negative score despite the considerable sorting skill shown by the close relation of mean outcome to forecast probability.

This example suggests that the skill score adopted by weather forecasters may not be ideal for solar forecasts of rare events. Some of the difficulty may stem from the use of a climatological forecast as a standard when solar climatology is in fact variable and difficult to predict. A more appropriate standard for flare forecasting might be a persistence forecast that uses today's outcome as tomorrow's forecast.

6.4 AVAILABLE EVALUATIONS

The program developed by Gray and Slutz (1967) has been used to evaluate SESC forecasts from 1967 to the present. Daily probability forecasts are compared to outcome in several ways. A plotted timeline, like that reproduced in Figure 6.1, shows each forecast and outcome in the context of data for surrounding days. Another graph shows mean outcome versus mean forecast for data grouped according to forecast. F score and climatology score are computed for each year and for each quarter. Despite the considerable sorting skill that we have seen in the SESC forecasts, the F score is typically worse than the climatology score. This is because of bias or unrealistic labeling--forecast probabilities are generally higher than outcome.

Jakimiec (1986) developed a new method of evaluating solar-flare forecasts. The distribution of deviations (difference between forecast and outcome) is described in terms of separate indices for large errors ("wings") and small errors ("core") and for asymmetry or bias. The distribution of deviations was found to change with the level of activity. The method was applied to a number of objective and subjective forecasts of flares and sunspots. More than half the tested forecasts were rated as "bad predictors" because of large wings or poor concentration to small deviation. The analysis gives insight into the sources of deviations but lacks the simplicity of the F score.

Heckman (1979b) discussed evaluation of SESC forecasts, particularly results of the validation study carried out by Gray and Slutz (1967) using F scores. Heckman redefined the terms "validity" and "sharpness" to mean, respectively, the ability to sort active regions according to the probability of burst production, and the ability to predict when the burst will occur, noting that solar forecasters can predict "where" with more skill than "when."

Heckman emphasized the effect on quality of forecasts of expending more effort and resources. Scores were compared for forecasts made by the part of the SESC group resident at the space center in Houston during the Skylab mission and forecasts made by the part of the group in Boulder. Individual forecasters commuted between locations, so the groups comprised the same individuals. Heckman attributed the higher scores achieved in Houston primarily to a more complete data base at Houston, and secondarily to "the high morale, concentration and interaction of the forecasters while in Houston."

SESC forecaster David Speich (telephone interview, 1984) corroborated this observation of increased forecasting skill for the SMM mission. He noted that the right region was selected and instruments were optimally pointed for 75% of the major flares for which experimenters controlled the pointing. Interpretation of this statistic is uncertain, but this experienced forecaster undoubtedly considered it to be an exceptionally good performance. In the first SMM mission, scientist-experimenters made the final decision on pointing, relying on information and advice provided by the forecasters, but no one has suggested that this involvement was a primary factor in forecast success.

Harold Zirin, director of Big Bear Observatory, wrote in a letter dated 1984 December 26, "We have succeeded over the years in catching almost all large flares that we could have seen from Big Bear. While we cannot predict the time of flares we can come close to 100% in picking the regions most likely to flare." J.B. Smith (1970) reported, in more quantitative terms, success in predicting flare locations.

One-day forecasts may give a poor reflection of forecasters' skills. Able to predict with confidence that a given active region is likely to flare within a 3- or 4-day period, forecasters deal with the temporal uncertainty by dividing the near-100% probability by the number of days (Heckman 1979b). This results in expression of a near certainty by an undistinguished forecast of 20-30%. A 3-day forecast might give a better opportunity for expression of skill. At the other extreme, forecasters believe that a "real-time" forecast on a much shorter time scale would also lead to improvement, presumably through the use of immediate precursors that are seldom useful in daily forecasts.

6.5 CONCLUSION

Practically, evaluation of forecasts at SESC, and supposedly at other centers, suffers from the same problems as data collection and archiving: forecasters are too busy trying to deal with the present and future to spend the time necessary to verify and record outcomes and compare them to their forecasts. Consequently, the forecaster's view of the future is based on a cloudy view of the past. Forecasting of rare events and the change of activity with the sunspot cycle introduce conflicting requirements: a long time is required to accumulate a sample of forecasts and occurrences of big bursts large enough to give valid results, but activity conditions change on a shorter time scale.

To summarize, predictive skill can be obscured by the exigencies of forecasting rarely occurring events. The F score seems to be most suitable for evaluating forecasts, but its application to solar forecasts needs to be examined. Before the skill score is readily interpretable, "climatology" or another standard needs to be defined for solar activity, and this requires data for the past. The value to the forecaster and user of proper evaluation of forecasts would be worth the considerable effort of carrying it out.

6.6 **REFERENCES**

Gray, T., and R. Slutz, 1967. A verification of the ESSA-Boulder Solar Disturbance Forecast Center forecasts. ESSA Tech. Memo. ERLTM-SDL 11.

Heckman, G., 1979a. Predictions of the Space Environment Services Center, pp. 322-349 in Solar-Terrestrial Predictions Proceedings, Vol. I: Prediction Group Reports, R.F. Donnelly (ed.). U.S. Dept. of Commerce, NOAA, ERL, Boulder, Colo.

Heckman, G., 1979b. Verification of solar flare forecasts at the Space Environment Services Center for the years 1969-1974. Unpublished manuscript.

Hunt, J.A., 1963. Decision theory and subjective probability in meteorological forecasts. M.S. thesis, Dept. of Elec. Eng., Massachusetts Institute of Technology (quoted by Sanders, 1966).

Jakimiec, M., 1986. Evaluation of the quality of solar flare predictions. Proc. of Meudon Solar-Terrestrial Predictions Workshop (1984), G. Heckman, M. Shea, and P. Simon (eds.), Boulder, Colo.

Murphy, A.H., and E.S. Epstein, 1967. A note on probability forecasting and "hedging." J. Appl. Meteor. **6**, 1002-1004.

Murphy, A.H., 1970. On the relationship between the "accuracy" and the "value" of probability forecasts.

Contrib. No. 191 from the Dept of Meteorology and Oceanography, Univ. of Michigan.

Neidig, D., P. Wiborg, P. Seagraves, J. Hirman, and W. Flowers, 1981, An objective method for forecasting solar flares, AFGL-TR-8-0026, Air Force Geophysics Lab., Hanscom AFB, MA 01731.

Sanders, F., 1963. On subjective probability forecasting. J. Appl. Meteorology, **2**, 191-201.

Sanders, F., 1966. The verification of probability forecasts. J. Appl. Meteorology **6**, 756-761.

Simon, P, 1979. The forecasting center of Meudon, France, pp. 1-11 in Solar-Terrestrial Predictions Proceedings, Vol. I: Prediction Group Reports, R.F. Donnelly (ed.). U.S. Dept. of Commerce, NOAA, ERL, Boulder, Colo.

Simon, P., J.B. Smith, Jr., Y. Ding, W. Flowers, Q. Guo, K. Harvey, R. Hedeman, S. Martin, S. McKenna-Lawlor, V. Lin, D. Neidig, V. Obridko, H. Dodson-Prince, D. Rust, D. Speich, A. Starr, and N. Stepanyan, 1979. Short-term solar activity predictions, pp. 287-321 in Solar-Terrestrial Predictions Proceedings, Vol. II: Working Group Reports and Reviews, R.F. Donnelly (ed.). U.S. Dept. of Commerce, NOAA, ERL, Boulder, Colo.

Smith, J.B., Jr., 1970. Predicting activity levels for specific locations within solar active regions. Presented as paper 10-1373 at the AIAA Observation and Prediction of Solar Activity Conf., Huntsville, Ala.

Vecchia, D.F., P.V. Tryon, G.A. Caldwell, and R.H. Jones, 1980. Statistical methods for solar flare probability forecasting. Air Force Geophysics Lab., Hanscom AFB, Mass.

7. Conclusion

7.1 SUMMARY

In this investigation of efforts related to solar-flare prediction we have looked at observational and theoretical studies of flare physics, at quests for flare precursors, and at mathematical models for combining masses of predictive information. We have also looked at the worldwide effort to gather and share timely data and combine it with knowledge and experience to forecast flares and their effects.

This endeavor requires an enormous variety of participants: dreamers and designers of telescopes and detectors, diplomats to arrange for cooperation, observers who can recognize and describe the relevant conditions and events on the sun, analysts and statisticians and theorists who can make sense of the observations, programmers and technicians to move the data and display it, and forecasters to put it together, finally.

Although no one can say exactly where the search to understand the flare process will lead, we can note progress: in recognizing that the energy must be released from a stressed magnetic field, and gaining quantitative understanding of some parts of the process; in seeing how energetic particles concentrate, transport, and distribute energy; in fitting together data from many sources to evaluate physical conditions; and in estimating the scale of the primary process and designing instruments to observe it.

More immediate gains for prediction are expected from certain practical efforts, if the experience of weather forecasters can be considered applicable to solar forecasting. If forecasts can be evaluated realistically, that is, in accordance with their value to the user, development and application of a method of evaluation is expected to focus the skill of the forecasters in useful directions. Application of numerical guidance, and development of the complete and accurate data base that it would require, is another step expected to provide the forecaster with a sharper view of the future by providing a clearer view of the past.

7.2 PLANS FOR THE FUTURE

The coming generation of experiments in space will continue to explore the solar spectrum, observing gamma rays, X rays, and radio waves. Multispectral observations from ground and space will be combined to gain new knowledge of the sun and flares. Instruments are planned to resolve spatial structures comparable to the scale of basic physical processes; perhaps the initial release of flare energy will be observed. Of interest for flare prediction is a proposal for a satellite in sun-synchronous orbit carrying continuously sunlit detectors to monitor solar radiation of geophysical significance.

NASA's major solar facility for the early 1990's is the Solar Optical Telescope (SOT). High spatial resolution, with enough sensitivity to acquire an image in a short time, is the goal of this effort to study the chromosphere, including flares, from the Space Shuttle. The resolving power planned for SOT is 0.1 arc sec, about 70 km, better by a factor of 30 than that of SMM. This will be a significant breakthrough to observe spatial scales comparable to those of physical processes in the chromosphere. [At the meeting of the Solar Physics Division of the American Astronomical Society in May, 1985, Alan Title showed video films of solar images made at the tower telescope at Sacramento Peak. These images, made from the ground as part of SOT preparations, attained spatial resolution of 0.5 arc sec.]

High spatial resolution and high sensitivity are attributes of the Pinhole/Occulter Facility (P/OF), another Shuttle experiment. A remote occulter will block light from the sun's disk and permit imaging of the corona close to the solar limb. A goal of P/OF is to provide powerful diagnostic information on particle acceleration and the site of energy release in flares.

Sunlab, part of the Shuttle's Spacelab, is a cluster of high-resolution detectors that include the High-Resolution Telescope/Spectrograph (HRTS) and the Solar Optical Universal Polarimeter (SOUP). Other experiments planned for Spacelab are Solar Active Region Observations (SAROS), the Spacelab-2 Hard X-ray Telescope, the Solar X-ray Polarimeter, and the Solar EUV Telescope-Spectrometer (SEUTS).

A white-light/Lyman-alpha coronagraph is underway for SPARTAN. Increasingly sensitive gamma-ray detectors will be flown. Meanwhile, the GOES satellites continue to monitor global X-ray flux, SMM already is well into its second mission, and plans for Hinotori-2 are underway. The new plans concentrate on exploitation of the enhanced spatial resolution now possible from space, on cooperative multispectral experiments, and on coordinated theory and analysis to maximize information return. The coordinating effort for the next maximum of solar activity, called Max '91, includes space experiments and supporting ground observations. Flare studies planned under this multi-institutional program include investigation of energy release in flares and determination of conditions for charged-particle acceleration.

A proposal of special interest to flare-forecasting is for a sun-synchronous satellite carrying continuously sunlit solar-radiation monitors. Instruments on the Solar Activity Monitoring Satellite (SAMSAT) are envisaged as a reliable source of X-ray and white-light coronal images, vector magnetic- field measurements, and global X-ray and EUV fluxes. Alternate proposed experiments, SAMEX and SXI, emphasize solar X-ray imaging with improved spatial and temporal resolution.

Back on Earth, Heckman at SESC has plans for progress. He anticipates making real-time forecasts and making more specific or detailed forecasts. The new computer should make the data more accessible. Heckman expects to add more optical data and new sensors and to get numerical guidance for forecasters.

The Very Large Array (VLA) of radio telescopes will continue to gather high-resolution polarization data from the ground, and the frequency-agile interferometer will exploit both spatial resolution and spectral versatility to measure magnetic fields in the corona and to detect flare precursors, perhaps useful for flare prediction.

Important for the future of flare forecasting is continuation and improvement of the primary data types, including X-ray bursts and H-alpha flares, magnetic class, and high-resolution observations of photospheric vector magnetic fields. H-alpha observations of chromospheric structures need to have the spatial resolution to show fibrils and the spectral versatility to show filament activations. Continued operation of the SOON sites (and addition of a fifth site in Italy) is planned, regardless of what monitors may be placed in space. The quantitative data on H-alpha flares from the SOON network should prove very valuable, especially if precise calibration between stations can be achieved.

Analysis of SOON data and comparison to X-ray burst data from GOES is a project that should improve the statistical knowledge of flare occurrence that has historically proved most useful in flare prediction.

APPENDIX A: GLOSSARY

ACRIM (Active Cavity Radiometer Irradiance Monitor)--Spaceborne instrument that measures variation as small as 0.1% in total solar irradiance.

ACTIVE LONGITUDE--Heliographic longitude range with abnormally high birth rate, growth, and persistence of active regions.

ACTIVE REGION (AR)--Plage, usually with sunspots. Surges, sprays, active-region filaments, and flares occur in active regions, and most radio bursts and noise storms can be associated with an active region.

ACTIVE-REGION SERIAL NUMBERS--See Hale active region number, SESC number, Mount Wilson sunspot number.

ACTIVITY COMPLEX--A cluster of active regions that often outlives the individual regions.

AFS: See Arch filament system.

AI: See Artificial intelligence.

ALERT--Warning issued from a forecast center to a specific user when a quantity of interest exceeds a pre-determined threshold.

ARCH FILAMENT SYSTEM (AFS)--A group of short, dark, curved features observed in H-alpha and interpreted as emerging magnetic flux.

ARTIFICIAL INTELLIGENCE (AI)--Computer programs that simulate human thinking using a series of decisions based on rules of thumb.

BETA: See Plasma Beta.

BOOLEAN COMBINATION--A combination of logical variables, such as "a or b and not c."

BRIER SCORE--The average squared difference between forecast and outcome. Forecasts close to reality have low scores. Considered a "proper score" in that it encourages the forecaster to make the most realistic forecast. Also known as F score.

BRIGHT POINTS (CORONAL X-RAY)--Brightenings that appear briefly and profusely away from active regions, but sometimes in association with the birth of bipolar magnetic regions. More X-ray bright points are detected near solar-cycle minimum.

BRIGHT POINTS (H-ALPHA AND EUV)--Small, temporary, flare-like brightenings in active centers or near active or potentially active filaments. H-alpha bright points are more numerous near solar-cycle maximum.

BRIGHTNESS TEMPERATURE--Temperature of a radio source that is directly proportional to flux density and to squared wavelength and inversely proportional to the solid angle subtended by the source.

BUMP-ON-TAIL--Feature of a velocity distribution that includes a distinct, high-velocity population large enough to cause a bump on the decreasing tail of the main distribution. On the low-velocity side of the bump, the number of particles increases as velocity increases, a situation conducive to instability.

Ca PLAGE, CaII K-LINE PLAGE--Plage observed in filtergrams made in light from a strong spectral line due to ionized calcium atoms.

CENTRAL-MERIDIAN DISTANCE--Heliographic longitude difference measured east or west from central meridian; it changes by the solar rotation rate of 13.6 degrees per day. See Heliographic longitude.

CLIMATOLOGY--Description of average conditions over a period considerably longer than the period between forecasts.

CLIMATOLOGY FORECAST--An average forecast that reflects only the occurrence frequency of the event forecast and not the specific conditions.

CLUSTER ANALYSIS--An informal method of numerical analysis that groups predictor variables in clusters that seem to be significant to the predictand.

COMPACT FLARE--Flare with small area and elevation, and often located at a foot of a single coronal loop. See Two-ribbon flare.

CORONAL GREEN LINE--Spectral line emitted by highly ionized iron (FeXIV) that indicates high density in the corona and temperature higher than the red line (FeX) but lower than the yellow line (CaXV).

CORONAL HOLES--Dark areas in imaged coronal emission: X rays, EUV, coronal emission lines, or microwaves. They are regions of low density and open magnetic field that underlie high-speed streams in the solar wind.

CORONAL MASS EJECTIONS--Outward-moving bright structures in the corona. Some are associated with flares and most with eruptive prominences or other chromospheric evidence of mass ejection. Their kinetic and gravitational energy is a large proportion of the total flare energy.

CORONAL SHOCK. See Shock.

CORONAL YELLOW LINE--Spectral line emitted by highly ionized calcium (CaXV), indicative of especially high temperatures in the corona.

DA (Discriminant Analysis)--A formal numerical method.

DARK FILAMENT: See Filament.

DELTA CONFIGURATION: See Mount Wilson sunspot classification.

DEPARTURE TIME--Launch time of a coronal mass ejection. Obtained by extrapolating to the surface the velocity observed in the corona.

DRIVER--A source of continued input of energy in a physical process such as magnetic reconnection.

EFR: See Emerging flux region.

ELECTRON-CYCLOTRON MASER--A process that transfers energy from trapped energetic charged particles to propagating radio waves related to the electron cyclotron frequency. (See Sec. 2.4.)

ELECTRON RUNAWAY--Condition that occurs when collisions become ineffective in inhibiting acceleration of fast electrons. This can happen when collisional loss of momentum decreases as electron speed increases.

ELEMENTARY BURSTS--Microwave bursts of very short duration, <20 ms. Correspondingly small source size and high temperature are deduced.

EMERGING FLUX REGION--A region where magnetic flux appears and grows, sometimes adding complexity and stress to the existing magnetic configuration.

EMISSION MEASURE--Measure of coronal emission lines. The product of the square of the electron density and the emitting volume.

ERUPTION--Description of prominences or filaments that ascend explosively. These include flare sprays and Disparition Brusque (DB), also called Sudden Disappearance.

EUV--Extreme ultraviolet. (See Fig. 2.1.)

F SCORE: See Brier score.

F STATISTIC--Ratio of variance between groups to the variance within groups. It measures the amount of variance of a dependent variable explained by grouping according to an independent variable.

FBS: See Flare buildup study.

FERMI ACCELERATION--Acceleration of charged particles at reflection in magnetic fields. First-order Fermi acceleration occurs when reflections that increase particle momentum dominate, as when the particle is trapped between approaching magnetic structures.

FIBRILS--Short, dark features seen in filtergrams, especially near filaments. Interpreted as lying parallel to magnetic field, and believed to display the direction of the transverse field and to reveal magnetic shear.

FILAMENT--Dark, very elongated feature seen on the disk in filtergrams, and appearing at the limb as a prominence. Filaments outline parts of lines of magnetic polarity inversion, or magnetic neutral lines.

FILAMENT DISAPPEARANCE--Transition to invisibility of a dark filament in a filtergram when the cool, dense gas of which it is composed becomes heated and ionized, or develops radial motion that Doppler shifts the absorption line out of the spectral window of the observing instrument, or rises into the corona and disperses. In general, a sudden disappearance would appear at the limb as an eruptive prominence.

FILTERGRAM--A solar image made in light of a narrow spectral window defined by a birefringent filter. Because of density and temperature stratification in the solar atmosphere, the wavelength range determines in which layer structures are observed. For example, at the center of a strong line of an abundant element, the radiation comes from the outside, or from the very top of the atmosphere.

FLARE BUILDUP STUDY--An international coordinated effort to observe and analyze preflare conditions on the Sun.

FLUX EMERGENCE: See Emerging flux region.

FOOTPOINTS--Intersection of a structure with a surface, as of the legs of a coronal loop with the photosphere or chromosphere.

FORCE-FREE FIELD--Field that exerts no force on the ambient plasma. In a force-free magnetic field, any electric currents flow parallel to the field, and any influence of the contained plasma is negligible.

GAMMA RAYS--Electromagnetic radiation at wavelengths shorter than 0.1 Angstrom and photon energy greater than 100 keV. are emitted in interactions involving highly energetic particles.

GEOALERT--A special coded message used by IUWDS to describe current and expected levels of solar and geomagnetic activity.

GEOMAGNETIC STORM--A worldwide disturbance of the geomagnetic field due to an abrupt change in the momentum and magnetic field in the impacting solar wind. Some geomagnetic storms can be related to interplanetary shocks identified with specific solar flares.

GOES (Geosynchronous Operational Environmental Satellite)--A series of NOAA satellites.

GRADUAL PHASE--A phase of slow change of flare emission, usually associated with processes occurring on a large spatial scale

GRE (Gamma-Ray Experiment)--An experiment on SMM.

GWC--Global Weather Central of the USAF.

H ALPHA--The spectral line due to transitions between levels 2 and 3 of the hydrogen atom. It is one of the strongest absorption lines in the visible solar spectrum. In flares, the absorption weakens, or reverses to emission, so that the flaring region appears bright in contrast to its surroundings. A variety of chromospheric structures are seen with high contrast in H-alpha filtergrams.

H-ALPHA FIBRILS: See Fibrils.

H-ALPHA FILAMENT: See Filament.

HALE ACTIVE-REGION NUMBER. Active-region number now assigned at Big Bear Observatory. This is a continuation of the plage numbers assigned at Mount Wilson Observatory from 1979 October through 1981 September and of McMath plage numbers assigned earlier at the McMath-Hulbert Observatory. Hale numbers in 1985 are between 10000 and 20000.

HALE BOUNDARY--An interplanetary magnetic sector boundary, or its projection on the solar disk, with magnetic polarities arranged in accordance with the sunspot polarity law described by astrophysicist George Ellery Hale: leader spots in the northern solar hemisphere have one polarity, follower spots the opposite polarity. Polarities are reversed for the leader and follower spots in the southern solar hemisphere, and polarities reverse from one activity cycle to the next. In cycle 21, with maximum in 1980, northern leader spots are positive. See Magnetic Sector Boundary.

HELIOGRAPHIC LONGITUDE--Longitude measured from the solar meridian that passed through the ascending node of the solar equator on the ecliptic on 1854 January 1 at Greenwich mean noon, Julian Day 2398220.0, reckoned from 0 to 360 degrees in the direction of rotation. This is the longitude tabulated in the Ephemeris and used in synoptic charts. Carrington's zero meridian passed the ascending node 12 hours earlier, so Carrington longitude differs from heliographic longitude, although the term is often used when "heliographic longitude" is intended. Longitude of a solar feature is nearly constant, unlike central-meridian distance, which changes at the solar rotation rate.

HELIUM I 10830 LINE--A spectral line with emission and absorption sensitive to coronal EUV radiation. It can be observed from the ground to provide information about radiation that cannot be observed from the ground.

HEURISTICS--Practical methods or rules of thumb, as opposed to procedures with a formal theoretical basis; literally "helping to discover or learn."

HINOTORI--The flare-observing satellite launched 1981 February 21 from Kagoshima Space Center, Japan. Interpretation of observations of X rays and gamma rays with spectral, temporal, and spatial resolution are contributing strongly to understanding solar flares. "Hinotori" means "sunbird, firebird, or phoenix."

HOMOLOGY--The quality of flares that occur successively at the same location, with similar characteristics. Flare homology has been cited as evidence that the magnetic field is little changed by the flare, or that it returns within hours to its preflare condition.

HOT THERMAL FLARE--A flare that is relatively strong in soft X rays, and relatively weak in hard X rays, interpreted as occurring in a preheated coronal plasma. (See Sec. 2.4.)

HRTS (High-Resolution Telescope/Spectrograph)--Instrument under development for SUNLAB on the Space Shuttle.

HXIS (Hard X-ray Imaging Spectrometer)--An SMM experiment that produced images of sources of X rays of various energies.

HXRT (Hard X-ray Telescope)--Telescope planned for Spacelab-2 on the Shuttle.

IMPORTANCE--Measure of flare size. (See Appendix B.)
IMPULSIVE PHASE--The abrupt increase in radiation, especially microwaves and hard X rays, that marks the transfer of energy from energetic charged particles to the surrounding plasma. (See Sec. 2.1.)
INFORMAL INTELLIGENT SYSTEMS--Another name for computer programs that store many facts and follow complex decision patterns rather than formal analytical methods.
INTERPLANETARY SHOCK: See Shock.
ISEE--International Sun-Earth Explorer.
IUWDS--International URSIgram and World Days Service.

KERNEL--A compact, bright knot in an X-ray or H-alpha image.
K LINE OF CaII--A strong chromospheric spectral line with complex line profile. It is the primary line for studying plages.

LABELING SKILL--The forecasting skill (complementary to the skill of sorting cases into groups with the same outcome) that correctly "labels" each group with a quantitatively accurate description of the outcome.
LANGMUIR WAVES--Propagating longitudinal electrostatic waves associated with plasma oscillations. In the corona and solar wind they may be excited at a frequency characteristic of the ambient plasma by a beam of energetic electrons.
LDE: See Long-Duration Events.
LOGISTIC REGRESSION--A formal mathematical method that determines a relation between a predicted quantity and predictor variables by minimizing a mathematical function that contains a logarithmic expression.
LONG-DURATION EVENTS--Soft X-ray bursts that decay slowly and may last hours. They are associated with flares or eruptive prominences of large area and elevation.
LOSS-CONE DISTRIBUTION--The anisotropic velocity distribution of a population of electrons trapped in a magnetic loop in which both density and magnetic field increase downward. Electrons moving parallel to the magnetic field reach low altitudes, collide in the dense ambient medium, and are lost from the loop population. The population remaining in the loop will be deficient in electrons traveling along the field.
LR (Logistic Regression)--A method of statistical analysis.

MAGNETIC ARCADE--A row of loops or arches forming a tunnel or gallery.

MAGNETIC BUBBLES--Closed 3-dimensional magnetic structures.

MAGNETIC COMPLEXITY--Departure from a simple bipolar arrangement; characterized by convoluted neutral line and occluded poles.

MAGNETIC ENERGY--Energy stored when a magnetic structure is sheared, twisted, stretched, or compressed, and released when the magnetic field takes on a configuration closer to that of a potential field.

MAGNETIC PLASMA--An ionized, magnetized gas.

MAGNETIC SECTOR BOUNDARY--Boundary where the prevailing direction of the magnetic field in the solar wind shifts from outward to inward or vice versa. Typically, 4 sector boundaries sweep by Earth during each solar rotation. Opposite magnetic polarities are envisaged as rooted in opposite (northern and southern) hemispheres of the sun, so that the dominant structure is a current sheet near the solar equatorial plane, separating opposite magnetic polarities. The polarity reversals occur where the equatorial current sheet is warped, so that it is located at higher latitude than the earth (or spacecraft) on one side and at lower latitude on the other side of the boundary.

MAGNETIC SECTOR STRUCTURE: See Magnetic sector boundary.

MAGNETIC SHEAR--Distortion of magnetic field due to relative motion of the plasma on opposite sides of a surface that intersects the magnetic lines of force.

MAJOR FLARES--Traditionally, H-alpha flares of importance 2 or greater; corresponds to X-ray class M2. (See Fig. 2.3.)

MAXIMUM ENTROPY METHOD--A formal mathematical method that selects the simplest result when incomplete data could be fit equally well by many different combinations.

McINTOSH SUNSPOT CLASS--Classification based on ZURICH evolutionary sunspot class. It also describes penumbra of largest spot and compactness of the sunspot group.

McMATH INTENSITY--Plage intensity.

McMATH PLAGE NUMBER: See Hale active-region number.

MICROFLARE--A small brightening, even smaller than a subflare.

MICROWAVE--The range of radio wavelengths extending from millimeters to decimeters

MIRROR POINT--Point at which a charged particle reverses the velocity component parallel to the magnetic field, where the field increases and field lines converge.

MODE INTERACTION--Interference or exchange of energy or momentum between two or more different modes of waves or oscillations.

MOS (Model Output Statistics)--A method used as numerical guidance in meteorological prediction.

MOUNT WILSON SUNSPOT CLASSIFICATION--Classification based on the magnetic configuration of the sunspot group:
- alpha: unipolar
- beta: simple bipolar group
- gamma: complex; polarities cannot be separated by a simple boundary
- delta: spots of opposite polarity appear within the same penumbra.

MOUNT WILSON SUNSPOT NUMBER--A serial number assigned to each sunspot group at Mount Wilson Observatory.

MOUSTACHE--Transitory emission wings of a spectral line arising from a spatially limited region; seen as bright points in a spectroheliogram or filtergram.

MSFC--NASA's Marshall Space Flight Center.

MULTIPLE REGRESSION--A formal mathematical method for expressing the dependence of a variable (the predictand) on known values of several predictor variables. A linear combination of predictor variables has coefficients chosen to minimize the squared values of the differences between predicted and actual values of the predictand.

MVDA (Multivariate Discriminant Analysis)--A formal method for developing a mathematical expression for quantitative prediction.

MVDA/CL (Multivariate Discriminant Analysis with Cooley and Lohnes procedure)--MVDA that does not assume uniformity of variance among the several variables.

NASA (National Aeronautics and Space Administration)--An agency of the U.S. federal government

NESDIS--National Environmental Satellite Data and Information Service.

NEUTRAL LINE--The boundary between line-of-sight magnetic fields of opposite polarity; also called line of polarity reversal. Non-zero transverse field is invoked to support the cool, dense gases in the filaments that often mark neutral lines.

NEUTRAL SHEET--A boundary surface between oppositely directed magnetic fields. Electric current flows in the neutral sheet, which is therefore also called a current sheet.

NGDC--National Geophysical Data Center.

NOAA (National Oceanic and Atmospheric Administration)--An agency of the U.S. Department of Commerce.

NOISE STORMS--Dense groups of brief, polarized, narrow-band, meter-wave radio bursts that may occur continuously for days. The source is in the corona over an active region.

NONTHERMAL ELECTRONS--Electrons with a velocity distribution in which one group of electrons is distinctive in the magnitude or direction of its motion.

NORMAL DISTRIBUTION--A symmetrical distribution of a variable characterized by a mean value and a variance. Also called Gaussian distribution.

OFF-BAND--Descriptive term for filtergrams or spectroheliograms that are made in light coming from a spectral band in the wings of a spectral line. They show Doppler shifted features and a deeper layer of the atmosphere than filtergrams taken nearer line center.

OSO (Orbiting Solar Observatory)--Satellite that measured X-ray flux and other quantities; followed SOLRAD and preceded GOES.

PCA: See Polar Cap Absorption.

PERSISTENCE--The tendency for high flare activity to persist for several days.

PERSISTENCE FORECAST--A forecast that present conditions will continue.

PITCH ANGLE--The angle between the trajectory of a gyrating charged particle and the direction of the magnetic field.

PIXEL--Picture element, one unit in a digital image.

PLAGE--An area of enhanced brightness in an active region, seen on a filtergram, especially H alpha or Calcium K. Plage corresponds to magnetic field that is stronger than average, although weaker than in sunspots.

PLASMA--Ionized gas.

PLASMA BETA--The ratio of thermal kinetic energy density to magnetic energy density.

PLASMOID--A package of magnetized plasma defined by closed lines of magnetic force.

P/OF (Pinhole/Occulter Facility)--A coronagraph planned for Sunlab.

POLAR CAP ABSORPTION--The enhanced absorption of radiowaves due to enhanced ionization of the polar-cap D region. The excess ionization is attributed to 10 MeV protons from the Sun, guided to Earth's polar caps by the geomagnetic field. Only the largest solar proton events cause PCA.

POWER LAW--A mathematical relation, $\log y = \log y_o - g\,x$, or $y = y_o x^{-g}$. The number of events of a given energy can often be described as a power law in energy.

PREHEATING--A relatively small rise in temperature that precedes some flares. (See Sec. 3.3.)

PRESTO--A message from a Regional Warning Center warning of the recent occurrence of significant solar or geomagnetic activity.

PROBABILITY FORECAST--A predicted probability or range of probabilities of the occurrence of an event (rather than a categorical forecast of occurrence or non-occurrence).

PROMINENCE ERUPTION: See Eruption.

PROTON FLARE--A flare associated with the detection of >10 MeV protons in the vicinity of Earth.

RECONNECTION--A process that simplifies magnetic structure and releases energy stored in the magnetic field. (See Sec. 2.3.)

REGION-DAY FORECAST--A forecast for a given active region for a given day.

REVERSED POLARITY--An arrangement of magnetic polarity in a sunspot group or complex opposite to the normal situation described by the Hale polarity laws.

RWC--Regional Warning Center of IUWDS.

RSTN--Radio Solar Telescope Network.

SAMSAT--Solar Activity Monitoring Satellite.

SAROS--Solar Active Region Observations (from Spacelab).

SECOND-PHASE ACCELERATION--A separate, delayed energization of charged particles postulated to explain long-lived, gradual radio and X-ray bursts that are delayed minutes after the impulsive phase.

SECOND-STEP ACCELERATION--Separate energization of electrons postulated to explain a delay of seconds of hard X-ray emission in some flares.

SELDADS (Space Environment Laboratory Data Acquisition and Display System)--System that will use a computer to store, retrieve, and display solar and geophysical data as numbers and graphs.

SELSIS (Space Environment Laboratory Solar Image System)--System that will display H-alpha images and longitudinal magnetograms received in the SESC forecast center within minutes of observation.

SESC (Space Environment Services Center)--Facility that

serves as IUWDS World Warning Agency and Regional Warning Center.

SESC NUMBER--Numbers assigned to active regions in which sunspots or flares occur or are judged likely to occur. The highest SESC number on 1985 May 01 is 4647, and may be called SESC 4647, AR 4647, or Boulder 4647.

SEUTS--Solar EUV Telescope-Spectrometer.

SHOCK--A disturbance moving faster than the phase speed of characteristic waves in a gas or plasma. Strong shocks have sharp gradients of density, temperature, speed, and magnetic field.

SHORT-TERM PREDICTION--A solar activity prediction made up to two days in advance.

SID (Sudden Ionospheric Disturbance)--A comprehensive term for various ionospheric effects of flare X-ray and EUV radiation.

SKILL SCORE--Measure of forecasting success beyond that expected from general knowledge such as occurrence frequency of the forecast event, or current conditions.

SKYLAB--An Earth-orbiting, sometimes staffed laboratory that studied the Sun from space in 1973-74.

SMM (Solar Maximum Mission)--Spacecraft launched in 1980, carrying a collection of experiments to study solar flare radiation from space.

SOLAR WIND--Continuous outflow of plasma and magnetic field from the Sun.

SOON--Solar Optical Observing Network.

SORTING SHARPNESS--The forecasting skill of sorting situations into groups with similar outcome. See Labeling skill.

SOT--Solar Optical Telescope.

SOUP--Solar Optical Universal Polarimeter.

SPACELAB--A set of experiments on the Space Shuttle.

SPARTAN (Shuttle Pointed Autonomous Research Tool for Astronomy)--Instrument that will observe the solar corona from space.

SPRAY--A kind of flare-associated eruptive prominence seen at the limb, with fragments that move outward along nearly straight trajectories and do not fall back to the Sun.

STP--Solar-terrestrial physics.

SUBFLARE: See Appendix B.

SUNLAB--A set of Space Shuttle experiments.

SXRP--Solar X-ray Polarimeter.

SURGE--An ejection seen at the solar limb of bright chromospheric material into the corona, usually associated with an observed flare. In most cases the material returns to the chromosphere along the same curved or straight path as it left.

SWF (Shortwave fadeout)--Fadeout caused by enhanced absorption in the ionospheric D region due to flare X rays.

SYNOPTIC CHART--Summary of activity or magnetic fields over a solar rotation.

TARGET REGION--The location where energetic particles collide and transfer energy to the ambient atmosphere.

TEARING-MODE INSTABILITY--One of several plasma instabilities that could contribute to the flare process; important because it does not require continued energy input.

TORSIONAL OSCILLATION--A large-scale, long-lived, cyclical pattern of surface velocities measured on the Sun. (See Sec. 2.2.)

TRANSITION REGION--A thin layer that separates the chromosphere from the corona. The temperature rises steeply outward from this region. The layer is geometrically convoluted by the presence of density structures such as spicules and prominences.

TWO-RIBBON FLARE--Flare that spreads along either side of a neutral line; associated with prominence eruption. (See Frontispiece.)

TYPE II (SLOW DRIFT) RADIO BURST--Radio emission that drifts with time toward lower frequency as expected if the source, believed to be a shock wave in the corona, moves toward lower density or upward in the corona. Typical velocity deduced from the frequency drift is 1000 km/s.

TYPE III (FAST DRIFT) RADIO BURST--Radio emission that drifts with time toward lower frequency at a rate corresponding to a speed of the order of one-third the speed of light; interpreted as plasma emission excited by the passage of a beam of relativistic electrons.

URSI--Union Radio Scientifique International.

URSIGRAM--Solar and geophysical data and advice, coded and distributed by wire.

UTILITY--A measure of the economic or other value of a forecast.

UV--Ultraviolet. (See Fig. 2.1.)

VARIANCE--A measure of scatter of values of a variable, usually defined in terms of the root mean square sum of residuals.

VECTOR MAGNETOGRAPH--An instrument that measures both longitudinal and transverse components of the magnetic field.

VLA (Very Large Array)--Radio antenna array that enables observation of astronomical objects with high spatial resolution (of the order of 0.1 arc sec in the microwave region).

WDC--World Data Center.

WWV--Radio transmitting station dedicated to time signals and solar and geomagnetic activity reports and predictions.

ZURICH CLASSIFICATION--A classification based on evolution of sunspot groups.

APPENDIX B: USEFUL QUANTITIES AND CONVERSIONS

LENGTH

Solar radius: 0.69599×10^9 m = 109 Earth radii,
subtends ~960 arc sec at Earth.
Astronomical unit (a.u.), half the major axis of Earth's
orbit: 1.4960×10^{11} m = 215 solar radii.
1 degree of heliographic longitude or latitude at center of
solar disk: 12,147 km.
1 arc sec is subtended by 725 km at disk center.
Limits of spatial resolution:
present limit from ground: <1 arc sec.
expected limit from instrument under construction,
operating in space: 0.06 arc sec.
1 Å (Angstrom) = 10^{-10}m = 0.1 nm (nanometer).
1 μm (micron): 10^{-6} m, 10^4 Å.

VELOCITY

Speed of light, c: 2.9979×10^8 m s^{-1}.
Thermal velocity: $(kT/m)^{\frac{1}{2}}$ cm s^{-1}. With Boltzmann's constant, $k = 1.381 \times 10^{-16}$, proton mass, $m = 1.67 \times 10^{-24}$, and temperature, $T = 2 \times 10^6$ K in solar corona, proton thermal velocity ~129 km s^{-1}.
Alfven speed: $B/(4\pi\rho)^{\frac{1}{2}}$. In the corona, at 1 solar radius above the photosphere, magnetic field B of 0.5 gauss and density, ρ, 0.9×10^{-17} g cm^{-3} give Alfven speed ~460 km s^{-1}.

MASS

Sun: 1.989×10^{33} g = 0.333×10^{6} mass of Earth.
Mass ejection from flare: 3×10^{16} g, comparable to mass of material ejected in eruption of Mount St. Helens in 1980.

FLARE SIZE

Limits of area defining importance of H-alpha flare:

Importance	Lower limit of flare area in:	
	Square degrees	Millionths of hemisphere
S	--	--
1	2.1	100
2	5.1	250
3	12.4	600
4	24.8	1200

(Largest sunspot recorded, 1947 April 7, had an area 5500 millionths of solar hemisphere.)

X-ray class	Lower limit of peak flux at Earth, in:	
	erg $cm^{-2}s^{-1}$	watts m^{-2}
A	10^{-5}	10^{-8}
B	10^{-4}	10^{-7}
C	10^{-3}	10^{-6}
M	10^{-2}	10^{-5}
X	10^{-1}	10^{-4}

An M3 burst has peak flux 3×10^{-5} watts m^{-2}; (the 3 in M3 is a multiplier, not the mantissa of a logarithm).

MAGNETIC FIELD

B, magnetic induction or magnetic flux density, measured in gauss; 1 gauss = 10^{-4} tesla.
Magnetic flux, measured in maxwells (Mx), is B (gauss) x area (cm^2).
Magnetic energy density: $B^2/(8\pi)$ erg cm^{-3}.

A moderately large sunspot (area 400 millionths of the solar hemisphere) has flux $\sim 10^{22}$ Mx; B ~ 3000 g at the center.
A typical magnetic flux element has flux of 6×10^{18} Mx.
In corona, at distances between 1.02 and 10 solar radii, $B(r) \sim 0.5(r-1)^{-3/2}$ gauss.

ENERGY FLUX

In a specific wavelength range, such as 1 to 8 Å, flux is measured in watts $m^{-2} = 10^3$ erg s^{-1} cm^{-2}.
Per unit range of frequency, f, F_f is measured in erg s^{-1} cm^{-2} Hz^{-1} (1 Hertz = 1 cycle per second).
Per unit range of wavelength, d, F_d is measured in erg s^{-1} $cm^{-2}\mu m^{-1}$ (1 μm = 10^{-6} m).
Because $f = c\, d^{-1}$ and $f\, F_f = d\, F_d$, $3 \times 10^{14}\, F_f = d^2\, F_d$.
At radio frequencies, flux is measured in watts m^{-2} Hz^{-1} or in solar flux units (sfu): $F(\text{sfu}) = 10^{22}\, F(\text{w m}^{-2}\, \text{Hz}^{-1})$.
Flux of X rays and gamma rays is measured in keV $cm^{-2}s^{-1}keV^{-1}$.
1 keV = 1.6022×10^{-9} erg; a range of 1 keV corresponds to a frequency range of 2.418×10^{17} Hz.
$T(K) = 4.799 \times 10^{-11}\, [f(Hz)] = 1.440\, [d(cm)]^{-1} = 11605$ eV.
Figure 2.2 shows a plot of the solar energy spectrum. The abscissa is log frequency, which can be converted to wavelength or to photon energy:

$$\begin{aligned} \log f(\text{Hz}) &= 10.48 - \log d(\text{cm}) \\ &= 14.48 - \log d(\mu\text{m}) \\ &= 18.48 - \log d(\text{Å}) \\ &= \log E(\text{eV}) + 14.39 \\ &= \log E(\text{keV}) + 17.39 \\ &= \log E(\text{erg}) + 26.18 \end{aligned}$$

The ordinate $\log (f\, F_f)$ is related to other flux units:

$$\begin{aligned} \log (f\, F_f) &= \log (d\, F_d) \\ &= \log F_d + 2 \log d - 14.48 \ (d \text{ in } \mu\text{m}) \\ &= \log F(\text{keV s}^{-1}\ \text{cm}^{-2}\ \text{keV}^{-1}) + 8.59 \\ &= \log F(\text{w m}^{-2},\ 1\text{-}8\ \text{Å}) - 14.5 \\ &= \log F(\text{w m}^{-2},\ 0.5\text{-}4\ \text{Å}) - 14.8. \end{aligned}$$

REFERENCES

Allen, C.W., Astrophysical Quantities, 3rd ed., Univ. of London, The Athlone Press, 1973.
Book, D.L., NRL Plasma Formulary, Naval Res. Lab., Washington, D.C., 1980.

Dulk, G., and D. McLean, Coronal magnetic fields, Solar Phys. **57**, 279-295, 1978.

SESC Glossary of Solar-Terrestrial Terms, U.S. Dept. of Commerce, NOAA, ERL, SEL, R/E/SE2, Boulder, Colo.

Solar-Geophysical Data, U.S. Dept. of Commerce, Boulder, Colo., February 1984.

NAME INDEX

SUBJECT INDEX

P

Q

R

S